CHRIS WALL
CUCKOO CORN

Tim Wait Publishing
2022

All rights reserved. No part of this publication may be reproduced, stored in a retrieval system, or transmitted in any form, or by any means (electronic, mechanical, photocopying, recording or otherwise) without the prior written permission of the author.

ISBN 978-1-3999-2765-9

A catalogue record for this book is available from the British Library.

Text copyright © The Estate of Chris Wait, 2022
Illustrations copyright © Tim Wait, 2022

Line drawings and front cover artwork by Tim Wait

Designed and typeset in Palatino by Charlie Webster,
Production Line, Minster Lovell, Oxford

Printed and bound by Short Run Press, Exeter, England

CHRIS WAIT
CUCKOO CORN

Contents

Introduction	7
January	9
February	22
March	36
April	49
May	59
June	73
July	87
August	102
September	115
October	130
November	143
December	158

Introduction

Twenty-five years after Chris's first year working on a small mixed farm in Surrey, he had witnessed enormous change. The horse had disappeared as a form of motive power, the combine replaced binding and threshing, hydraulics and electricity replaced manual toil and skills and everything got bigger – fields, gateways, buildings, tractors and implements.

All of this was inevitable as British farmers rose to the challenge of feeding a hungry post-war world. An unfortunate consequence of this push for higher-yielding crops was arable farmers becoming increasingly dependent on the chemical industry and their products. Environmental scientists are now recognising the damage that chemical fertilizers have done to the environment and have started a countdown to the day when our land will no longer be able to sustain crops. Chris recognised the beginnings of this trend and wrote articles about it, speaking as 'someone who doesn't want to end their farming career cultivating the ground with a deer's antler'.

In the preceding twenty-five years Chris had married, done National Service as a policeman in the RAF, run a small tenant farm on Sark, been head herdsman to Lord Mountbatten, and managed farms in Kent, Sussex and Wiltshire before taking on four farms as a single thousand-acre unit on the Oxon-Bucks border.

This varied experience led Chris to question the validity of modern farming practices. His musings led to the start of his writing career: first letters and articles, then some short stories and finally *Cuckoo Corn* which he completed in 1985.

Sadly Chris died of heart problems two years later and never got to see his book published.

Thankfully there is now an upsurge of interest in the environment and also how we manage the countryside. I believe Chris's book provides a valuable insight into how farming used to operate. He saw a sympathy and respect for nature and natural forces rather than nature being a force to be overcome, bent to our wills and dominated. He also makes it a very entertaining read.

Chris's school days were spent in an ancient boarding school where the ancient traditions – fagging, bullying and general savagery – were upheld and he detested it. During his summer holidays, Chris was drawn towards the low grinding drone which indicated a Fordson tractor was engaged in harvesting work nearby. He overcame his shyness and joined in with the work. The honesty and great humour of the farm labourers created in Chris a lasting love and respect for these hard-working, knowledgeable and highly skilled people, an antidote to the future bank managers and politicians with whom he was incarcerated at school. Farming became his refuge.

Chris would be happy that *Cuckoo Corn* is finally in print but nothing would delight him more than to be playing a part in the increasingly important debate as to how we are to avoid going back to scratching the ground with a deer's antler.

Tim Wait

1

JANUARY

Cold toes, cold fingers, and new rubber boots slipping on the icy pavement. Hoar frost shining white on the paling fence beside the watercress beds and deep shadows in the farm entrance contrasting with the moonlight of the yard where it was bright enough to read. I remember the details of the short walk to work on that frosty Monday morning in January 1950 because it was an important occasion.

I had just left the close confines of a boarding school to start a career in farming, and that Monday was my first day as a full-time farm worker on a small, mixed farm at Abinger, in Surrey. It was a big step from being a boy who helped on the farm during school holidays, to being 'Boy' on the same farm. Overnight I had changed from being the 'whining schoolboy…creeping like snail unwillingly to school', to the next step in Shakespeare's seven ages of man – the lover. Not a lover of ladies, I hasten to add – after seven years at a public school where one was taught that the fair sex were the descendants of Eve, sent into this world to seduce the thoughts of young men away from rugger, cricket, the Empire and a studious and rewarding career in Daddy's bank – I was terrified of women. But I was in love with farming and in particular, the farming that was in progress at Hatch Farm, Abinger Hammer.

Hatch Farm was a typical Surrey farm of the early Fifties. The land – about 120 acres – ran from Hackhurst Downs in the north, down across the railway to the A25 Dorking to Guilford road where the farmhouse and buildings stood. They still stand there, although now somewhat forlornly, having lost the bustle of being the hub of an active, working farm. Then across the Goose Green on which we had grazing rights to the steep hill of the Hangers, which was the start of the main body of the land, stretching along the Holmbury road as far as Fulven's Lane, the southernmost boundary.

The soil, with the exception of four acres of heavy clay next to the railway, and the chalk under the Downs, is the free-draining and easily worked Surrey sand. This sand needs constant topping-up with organic matter, either by grazing animals, the dung cart, or by ploughing in green crops. But given sufficient rainfall, it will produce heavy crops of grass and hay, and the straw crops which, by the standards of the Fifties, were very respectable indeed: up to 35 hundredweights to the acre, or as we said, fifteen sacks to the acre, when the national average in the 'bumper harvest' year of 1949 was only 22½ hundredweights. So the ground was ideally suited to the mixed farming regime carried on at the time.

The main enterprise on the farm was the herd of seventeen Shorthorn cows, milked by hand in the old wooden, lean-to cowstall at the back of the yard. All the calves were reared, the heifers for replacements and to increase the size of the herd, and the best of the bull calves to be steered and run on with the heifers until they were sold off as store cattle. There were three or four breeding sows producing weaners for sale at about eight weeks old, and the boss would keep back a likely-looking litter to be taken on to pork weight, fed on swill collected from the Abinger Arms, the pub just up the road, next to the Hammer Clock. We boiled the swill up in a big, brick-built copper next to the calfyard.

Apart from the few dozen chickens running 'free range' in the paddock next to the farmhouse, the only other animals on the farm were the two heavy horses. They provided half the motive power on the land, the other half being an asthmatic and ancient Fordson tractor which, once you had persuaded it to start on a cold morning, gulped paraffin at an alarming rate, burnt a gallon of engine oil each day and wept large quantities of water and gear oil as it groaned its way across the fields.

I had none of the usual qualms about meeting my new boss or the men I would be working with as I walked to work on my first day as a 'regular'. For the previous two years I had spent every day of the school holidays at the farm. At first I was ignored and had to be content with being a longing spectator, but I persevered and gradually progressed from being allowed to ride on the tractor wing to actually steering the thing as it chugged its way between the stooks in the harvest field. And from the lowly shovelling of muck in the cowstall to the Eleven Plus of being pronounced competent to milk Old Daisy, the quietest and easiest milker in the whole herd.

"Make summat o' you, Boy, one day we will. If we doesn't die o' old age fust, that is."

The boss was a farmer's son with generations of farming behind him. He joined the Surrey Yeomanry before 1939 and had been on active service throughout the war, coming out through Dunkirk and fighting in North Africa and Italy. Now that the war was ended he threw himself back into farming with tremendous energy and enthusiasm. Altogether the ideal person to teach the job to a green youngster. And most important of all – from my point of view – he had a variety of enterprises on the farm and was starting from scratch with very limited capital. There were few labour-saving devices so I was taught the basic skills the hard way, by hand.

The first of these jobs each day was, of course, the milking. Being 'Boy' I was expected to arrive first in the yard and get things ready for the experts to swing into action. Ernie Sayers, the cowman; Fred Bartlett, the carter; and Bob Newman, the boss.

There was no electricity in any of the farm buildings so the first thing to do was get some light. A 'Tilley' pressure lamp was left filled and left ready for the morning on the table. I fumbled about in the moonlight, found the little scissor-like padded clamp, dipped it in methylated spirit from the jam jar on the table, and then clipped it on the central vaporising tube under the mantle of the lamp. It burned with an eerie blue flame and seemed to give no heat at all, but very soon the lamp started to hiss and pop as the paraffin in the tube started to vaporise. I waited for it to warm up properly and then gingerly pumped up the reservoir until the lamp was giving off a steady hiss and a brilliant white light. It had to be done carefully, because if the lamp was pumped too hard, or too early, it would turn into a miniature flame-thrower and send a roar of smoky, yellow and orange flame up to blacken the ceiling. Then Ernie would have been 'real spiteful'. He was very particular about the cleanliness of his dairy.

Next, the dairy boiler – also left ready for lighting with the water jacket filled and the whole thing draped in sacks to protect it against frost. I took the sacks off, checked the water level in the sightglass on the side, opened the firebox door and pushed some newspaper under the dry sticks that had been left there overnight. A quick fumble with cold fingers and the matchbox again, the firebox was slammed shut and up she roared. It drew well in the frosty air, that boiler, and soon there'd be red embers dropping through the grate and I could fill it up with coke. It only took a few minutes until the water was warm enough to wash the cows' udders, and by the time milking was finished there'd be plenty of boiling water to scrub the buckets and the cooler, and there'd be a good head of steam for the old-fashioned sterilizing chest.

It was easy, at that time of day, to squat in front of the

boiler staring into the cheerful glow and fall into a doze. Some farms had electric boilers but I liked ours. You can't warm yourself in front of an efficient, heavily insulated 'Electrobloc'. A roaring fire has a mesmeric effect on a cold morning which is virtually impossible to resist…but my daydreams were interrupted by brisk, energetic footsteps in the yard. Ernie, the cowman, marching in at the end of the two-mile walk from his council house in Gomshall.

"Mornin', Boy. Got the fire goin' then? Done them gutters yet? Come on – what you been at? Thought you'd 'ave milked 'alf on 'em by now! Gorblimey! Let's get started, we han't got all day."

And Ernie bustled about getting milking buckets out of the sterilizing chest where they'd stood upside down to drain overnight.

I took the lamp and left the dairy lit by the glow of the boiler while I went into the long cowstall to tidy the bedding ready for milking. As I did so there was a great heaving and stretching, a rattling of neck-chains and rustling of straw as the seventeen Shorthorns got to their feet when they heard the door open. They turned their heads and watched me walk behind them, ears forward and big, mild eyes shining and blinking in the sudden bright light.

During the night some of the bedding had been pushed back into the gutter. I went along the standings with a prong and forked any straw that wasn't soiled with dung back up around the cows' forefeet. The rest, and any dung that was up on the standings, was scraped to the back of the gutter. The milking stool and bucket mustn't be allowed to get dirty. Ernie had emphasised the importance of cleanliness when he first taught me to milk in the school holidays.

"You doesn't just sit down to a cow any old 'ow, Boy, you does it proper. First wash your 'ands. Then make sure the stool an' bucket's clean 'cause you got ter pick 'em up wi' yer milkin' 'ands. Stool in the left 'and, bucket in the right – like so. Put the stool under yer backside as yer goes down an' 'old the bucket atween yer knees, off the ground so's it don't clank an' make yer cow jump."

He made me sit down to an imaginary cow in an empty standing a few times until I'd got it right. Then, and only then, was I allowed to try my luck with a real cow – old Daisy, of course – but still with the constant of flow of instructions.

"Stand up alongside yer cow an' then talk to 'er. Let 'er know you're a'comin'. The sit down wi' yer 'ead against 'er flank nice an' smooth an' easy, like. Come at 'em a bit sudden, or startle 'em, an' they'll kick you into next week, more'n likely. But you go at it a bit gentle an' you'll be all right. Keep on talkin' to 'em, that's the secret. They likes the sound o' yer voice."

The soothing human voice is a valuable and potent tool in the business of animal husbandry. All good stockmen, whether they be in charge of cattle, sheep, pigs, horses, or even poultry, talk to their charges incessantly. The trouble is that speaking one's thoughts aloud becomes a habit. This leads to confusion in the minds of ignorant townspeople – they see Hodge carrying on an amiable conversation with no-one in particular as he goes about his business, and dismiss him as the village idiot. In reality he is probably a highly-skilled craftsman doing his job 'a bit proper, like'.

When I'd finished tidying the standings Ernie came into the cowstall carrying his milking bucket and stool, and the old galvanised bucket we used for udder washing. I washed the teats of the cow he was going to start on and he sat down to her. Soon the sound of milk pinging into the empty bucket changed to a frothy, swishing sound, each squirt blending into the next so that it sounded continuous. Watching an expert like Ernie milk, you'd scarcely see any movement in his arms except the small muscles in the forearms rippling as his hands opened and closed.

The first cows to be milked in the mornings, and the last at night, were the recently-calved high yielders. This was so that their milking interval would be as close as possible to the ideal twelve hours. I wasn't allowed to milk the 'heavies' until I'd got more experience on the old, quiet cows near the end of their lactations, but there was plenty to do. I stood by ready to take a full bucket from the cowman and give him an empty one so that his milking shouldn't be interrupted. A cow giving seven gallons a day gives about four at the morning milking, and you can only get two and a half gallons into a milking bucket without the froth spilling down the side.

I carried the full buckets carefully along the passage to the dairy and tipped the warm milk into the big pan over the cooler. Then I turned the tap on just enough for the milk to trickle over the cooling surface. This was like a radiator except that it had cold water and not hot running through it. At the bottom of the cooler the milk ran into a strainer with cotton-wool filters in it, standing on top of a ten-gallon churn. It was my responsibility to change the churns as they filled, and woe betide me if I forgot and one overflowed onto the dairy floor.

Soon after milking started the farm cats appeared in the stall. You never saw them creep in, they were just suddenly there. They crouched, immobile, opposite the milker, watching him with unblinking eyes. They were half wild and wouldn't accept a stroking hand. All they wanted was a dish of warm milk, which they took as their right without any of the displays of purring gratitude shown by soft, domestic puss. They lapped the milk slowly and with concentration, staring suspiciously and quickly round at any sudden movement in the cowstall while they were drinking. When the dish was licked clean they washed the drops of milk from whiskers and chin and vanished just as quickly as they had come, about their own secret business, to reappear just as suddenly again at the next milking. They never got food from us, just a lick of milk, but they were always sleek and healthy, close-furred and lean; deadly efficient hunters engaged in the constant and stealthy war against rats and mice.

The team of cats was an important part of the farm economy. Rats and mice, if allowed to run unchecked, cause an enormous amount of damage to stored grain, feeding-stuffs, and the sacks and bags in which fertilizers are stored. They carry disease to humans and animals and their constant gnawing and burrowing seriously damage farm buildings. We never had to resort to expensive and dangerous programmes of poisoning – the cats did the job for us. But we had to cooperate with them. Every doorway had a cat-hole cut into it so that the cats could patrol all parts of every building, and sacks of grain and feed were stacked in lines. Six-inch gaps between the lines ensured that there were no refuges for vermin which the cats couldn't reach.

It was vitally important, too, that the cats weren't fed but just given milk. At Paddington Farm, just up the Dorking road, where the boss's mother farmed, the cats were treated as pets and liberally fed. I watched them dine one evening when I was sent up there on an errand. They were fat and happy and there were seventeen of them. The farm was overrun with rats.

At seven o'clock the boss brought a jug of tea down to the stall. It was hot and sweet with plenty of milk in it; or if he was short of milk in the farmhouse kitchen, the boss held the jug under the nearest cow and milked a few squirts into it. The tea frothed up and retained its head when poured into a cup, but it was warming and tasted good.

Drawn by the smell of tea, old Fred, the carter, came from the stable where he had been seeing to his horses. Fred didn't have so far to walk in the mornings as Ernie did. He lived in one of the farm cottages opposite the little grocery shop at 'Frogs' Island', a group of cottages where most of the men employed in the village's main business, the watercress beds, lived. So Fred only had to plod a hundred yards or so to work in the heavy, hobnailed boots he always wore.

Fred was a real craftsman. He's worked at Hatch Farm since long before the war but came originally from Sussex. The Sussex burr was so thick in his speech that it was some weeks before one could, with practice, decipher the words he used. But it was worth making the effort. They were words of wisdom, culled from sixty years of living on the land and taking a pride in good craftsmanship. Very early on Fred helped me get my priorities right.

"Allus 'member you be workin' fer the farm, Boy, an' not the man." he said. "Bosses do come an' go, an' so does blokes, but the farm do go on."

When we had gulped our tea we got on with the milking. With four of us at it, it didn't take long and the time passed quickly with the interesting, shouted conversation that ran up and down the long cowstall. The weather, plans for the coming day, remarks about the previous day's work and the condition of animals and crops, comments on the performance of local football teams, and pungent details of items of village gossip picked up in the bar of the pub the night before.

Villages had a very efficient bush telegraph in those days and morning milking was an essential part of it. A moral peccadillo committed in the afternoon was reported, suitably embroidered, in the pub the same evening. The matter was then discussed in detail in cowstalls within a three-mile radius of the village at first light. It was passed on to milk-lorry drivers, postmen, policemen, and any other early birds who cared to listen, by about eight o'clock the following morning – and, of course, the tales lost nothing in the telling. Quite often one would listen enthralled to some particularly spicy story and realise, after the speaker had been going on for time, that it was the same tale you had told him – in strict confidence, naturally – some twelve hours before.

When we'd finished milking, Fred went to feed the pigs while the boss and I fed the baby calves. They always had their own mother's milk for the first three days of life. The thick, yellow colostrum -'beastings' as Ernie always called it – contains antibodies and vitamins which the new-born calf lacks. Without this natural vaccination from the mother, the calf would quickly succumb to one of the common infections like pneumonia or scouring which an established animal's immune system guards it against. And the colostrum has a high level of protein and forms soft, easily-digested curd in the calf's rumen. This gives it a good start.

Getting a new-born calf to drink from a bucket is a tedious and sometimes infuriating business. The calf's natural instinct is to nuzzle upwards for a teat. To persuade the little animal that gravity generally makes milk fall to the bottom of a bucket takes time and patience. And another calf's instinct which is not helpful is that of butting upwards at the mother's udder to make her let her milk down.

The usual procedure is to back the calf into a corner of the pen, stand astride it, hold the bucket of warm milk in one hand and get the calf to suck the fingers of the other hand. The calf's mouth is then drawn down into the milk, it keeps on sucking and automatically starts drinking. The fingers can then be withdrawn from its mouth. It's delightfully simple in theory.

What often happens in practice is that the calf refuses to allow its head to be drawn *down*, when its instinct tells it that the milk is somewhere *up*. It can smell the milk and knows that it's going to be fed, but so far the surrogate teat it's been sucking vigorously hasn't provided any milk. So it butts with its head to remind Mum to deliver the goods. Its sharp little incisor teeth cut the fingers that are in its mouth;

its hard little head bruises the knuckles; and the milk in the bucket is projected violently upwards and outwards.

Half a gallon of warm, sticky colostrum in your hair, down the front of your shirt and trousers and trickling gently into your wellingtons is not conducive to good relations between man and animal.

But when the battle of the bucket has been fought and won, feeding the calves was a straightforward business. It was just a matter of remembering which ones were on colostrum (and who their mothers were); which ones were on whole milk for the first ten days and which ones had graduated to 'Spillers' Calf Gruel' – a mysterious, grey-brown meal which, when mixed with hot water from the dairy boiler, resembled a thin and gritty porridge.

Ernie had been busy in the dairy while we were feeding the calves. He rinsed all the buckets and the cooler with cold water to prevent 'milk stone' forming on the metal, and left them to soak in the big washing-up trough. They would be scrubbed with hot water and detergent after breakfast, before being stacked in the sterilizing chest to be sterilized with steam under pressure for half an hour or so. Then he ladled milk in or out of each churn until the level was exactly on the gallonage marks stamped into the sides. The labels were filled in with our name and address, the amount in each churn, and the number of gallons sent to the dairy in Westcott that day. The final rite in the whole ceremony of milking – the blessing, so to speak – was to tuck a label under each churn lid and enter the total in the grubby notebook hanging in the dairy. The milk was then out of our hands and became the property of the Milk Marketing Board.

When we carried the churns out across the yard to the churn stand beside the road I complained about the 150 lbs weight of each full, ten-gallon one. Ernie soon corrected me.

"Bliddy 'ell, Boy, you got it easy, nowadays, you 'ave. Afore the war we used ter 'ave them seventeen-gallon railway churns. Over two 'undredweight they was. Took some liftin' onto the cart they did – 'cause in them days you used ter 'ave ter take yer milk ter the nearest railway station. Took more'n a skinny young booger like you to lift 'em."

If Fred had been there he would have said, 'Wants to eat more pudden, Boy.' He always said that when I grunted awkwardly under a heavy load – not having acquired the farmworker's knack of lifting with economy of effort and without straining myself, but the boss grinned and said,

"He'll have to wait 'til we're thrashing next winter, won't he, Ernie. We'll show him what lifting's like then."

"Ar," said Ernie, "time 'e's run up an' down them granary steps wi' a two-an'-a-quarter sack o' wheat on 'is back a time or two, 'e'll know all about it."

"He'll know all about it." The words ran through my mind as I walked home to breakfast. Maybe by the time thrashing came round I might, indeed, have got the knack of some of the jobs on the farm; but I had no illusions about the length of time it would take for me to become a fully-trained farmworker. Wise old Fred had put me straight on that score right at the start. My first attempt at ploughing during the school holidays had turned out rather well – the boss had set the plough for me – and I had bragged to Fred about it.

"I seen wuss," he admitted. "if I cast me mind back a good number o' years I can just 'member seein' wuss'n that."

This was high praise from Fred and I glowed with pride. But Fred had more to say:

"But you wun't win no ploughin' matches wi' work like that. You got ter 'ang on a bit 'til you've 'ad more 'sperience afore you starts shoutin'."

"How long will that take, Fred?" I was thinking in terms of weeks or months.

"Oh, not long, Boy," he replied. "if you works at it you should be summat of a ploughman in ten year or so."

At sixteen I was still measuring time by school terms which dragged on for a lifetime. Ernie and the boss had been talking casually about thrashing next winter. I relished the thought that before that time came I had the equivalent of three terms and the holidays between them, of exciting revelations and experiences. And after that? If next winter was far away – ten years time was light-years distant in the misty future. I pushed away thoughts of what was in store for me and concentrated on what the present had to offer.

After breakfast I might be helping Ernie with his routine work in the yard and cowstall, or go with Fred, carting in fodder for the cows. And if I was extremely lucky it might be one of those days when I helped the boss do something exciting with the tractor.

It is the small changes in routine and the variety of the job that makes farming such an interesting and challenging occupation. The first things you learn are that in a constantly changing business ruled by the vagaries of the weather, animals and market trends, you must be prepared to learn new things every day of your life. And that in order to survive you need to pitch in and do your best, getting fun and enjoyment out of the job in return. I was excited at the prospect of learning from experts, and happy in the knowledge that the learning process would be full of fun. The three men I would be working with had the right, light-hearted approach and even petty squabbles and disagreements were *always* laughed about afterwards.

So my frame of mind at the beginning of that first day at work was one of contentment and excited anticipation. A strange combination but the right one for a youngster with the world at his feet.

2

FEBRUARY

February on the farm seems to go on for ever. The land is held hard in the grip of winter and the wind blows cold and unceasingly from the east. Humans and animals shrink away from it as they endure the hardship; milk yields dip down and store cattle stand with lowered heads. However much fodder is stuffed into the animals, they don't grow – it takes all they can eat to maintain them and keep them warm.

Being 'Boy' on the farm there was no set pattern to my work – apart from helping Ernie with the morning milking. I was sent to help where my willing but inexperienced hands could do the most good – or as Ernie constantly reminded me, where I could do the least harm. This was good for me as not only was there variety, but I was shown the right way to do the hundreds of small jobs that go to make up competent farm work – each small job having a knack of its own which has to be learnt.

Ernie's and Fred's methods of instruction were different. Ernie explained in great detail how a job was to be done, and then make me practise each stage until I'd got it right. He didn't quite shout, 'Hon the command – ONE!', but I'm sure he would have done if he'd ever had any army training.

Fred, on the other hand, allowed me to fumble my way into a task – doing it all wrong – then he'd put me right. I remember the first time we went together to get a load of straw from a rick. Fred leant on his prong with an amused twinkle in his eye, his jaws moving rhythmically as he chewed his tobacco (he never smoked near a rick but had to have his 'bacca' all the time), while I struggled for a couple of minutes, trying to lift a truss that seemed inextricably bound in with the mass of the rick. Then he stepped across.

"You goo on like that, Boy, an' you'll break the bliddy prong 'andle. 'Tisn't no good a-tryin' ter lift trusses what be pinned down wi' t'others – an' you'll never move that booger there 'cause you be standin' on t'other end of un. Take 'em out a bit sensible, like. They went in one atop've t'other, didn't 'em? So they comes out back-end fust. Last in, fust out – that's the way."

And I quickly got the hang of it.

Like all craftsmen, long before the days of craftsmen's certificates, Fred was a time-and-motion expert. He showed me the easy way of carrying out each job, not because he was lazy, but because if you take a steady pace and do the job correctly you achieve much more in the day without becoming exhausted – like his insistence that I learn to use a prong right-handed as well as left-handed, making those awkward corners in the dung-yard or on top of a rick so much easier to get at.

During the winter Fred and I spent a lot of our time fetching hay, mangolds, kale and straw in from the fields to feed the cows. Kale was the worst job, it had to be cut daily and fed before it wilted. Often the plants were covered in frost and Fred went along with a branch cut from the hedge and brushed the ice off. Then we bent our backs in the freezing wind and cut it off close to the ground with faghooks. After a few minutes of hacking away at the hard-frozen, three-inch thick stems of the Marrow-Stem kale my fingers went numb and I found it hard to control the wet and slippery handle of the hook.

"You goo stiddy, Boy, you'll chop your bliddy leg off if you whangs around like that."

Fred's gnarled red hands seemed impervious to the cold as he sliced skilfully at the kale, the corn sack tied round his waist keeping his legs dry, and the one over his shoulders keeping the rain and sleet off.

Mangolds weren't so bad – they were, at least, dry as they lay in their snug clamp, protected against frost by a thick layer of straw and earth. All we had to do was throw them up into the cart, and it didn't take long to get a load as each mangold was the size and shape of a rugger ball.

I'd been dipping into my new copy of 'Watson and Moore' and thought I'd try some of the new-found knowledge out on Fred.

"Did you know that mangolds are 90% water, Fred?" I asked casually.

There was a grunt from the other side of the cart and another couple of mangolds sailed up and over and thudded onto the load.

"If mangolds are mostly water," I persisted, "why do we bother to grow them? Why not give the cows water out of the tap?"

This time there was a snort, then Fred's head appeared over the side of the cart.

"What the 'ell you on about now, Boy?" he demanded, then comprehension dawned. "Ar, you been book-learnin' again, I can see that. Now, what you mean, all water?"

"That's what the book said. Why do we bother to grow them?"

Fred picked up a big mangold and hefted it in his hands. It weighed about ten pounds.

"If I were to bash you over the 'ead wi' this, Boy, you'd reckon as 'twere more'n just a bit o' water," he said, "an' that's what they be. Cows do milk real well on 'em – they likes a bit o' summat juicy an' sweet when they han't got no grass. You try givin' 'em a bucket o' water 'stead o' their mangolds next time you be feedin' 'em along of Ernie an' 'e'll give you a book-learnin', 'e will."

"I only asked, Fred. The book said…"

He interrupted me. "An' if us don't get this 'ere load o' water down 'ome a bit quick, boss'll 'ave summat ter say ter we. An' 'is words won't be out o' no book neither."

On the way back down to the yard Fred relented.

"I knows as 'ow you likes lookin' at them books o' yourn, Boy, but you 'member you got to learn the job practical fust. Why, that's what you're 'ere wi' us for, an't it? Afore you goes ter college? We got ter learn you a bit proper, like, so's you can put they college teachers right when you gets there."

"Didn't you ever go to school, Fred?" I asked.

"No, Boy, I went straight on the farm. Bird-scarin' at a shillin' an' a tanner a week, I was. Never 'ad no schoolin' at all." He thought for a while then grinned at me.

"Now you knows why I 'as ter make up for it by bein' a bit bliddy artful."

In those days we didn't have any fast, efficient means of transporting fodder to the yard directly it was gathered; we only had horses and wagons or the lumbering old Fordson. Hay was stacked in the field in which it grew and the stack was then securely thatched so it didn't need a barn. Hauling it home was left until winter when there was more time available.

When we went to get hay, Fred got one of the horses out and put her in the shafts of the two-wheeled dung cart. The ladders were laid in the cart as you never put them up in an empty cart; they sway and rattle about and get broken. We collected the wagon rope from its peg in the stable and made sure we had the 'rubber' – the carborundum sharpening stone. The big hayknife, used to cut hay out of the rick was left up there from last time, covered in old thatch to protect it from rust. You left tools in the fields in those days and no-one would dream of stealing them.

Up deeply-rutted 'Hollow Lane' to the top of the Hangers we plodded. I led the mare and Fred, with both hands stuffed into the pockets of his old army greatcoat, walked behind the cart. We never rode in a cart in winter, it was better to walk and stay warm. And an empty cart rattles and bumps uncomfortably on frozen ground.

There wasn't much room to walk in the lane leading a horse – just a narrow ridge of earth between the wheelruts and the central track where the horses trod.

Farm tracks are different now. They have shallower, wider ruts because rubber tyres don't cut in as much as cartwheels did, and the centre is overgrown with weeds. After hundreds of years the ridged, central path of the horses has disappeared, although if you look carefully at the old droves on the tops of the Downs in Hampshire and Wiltshire, it is still faintly discernible.

We got to the hayrick at the top of the Hangers, set the ladders up in the cart, threw out the rope, and while Fred sharpened the hayknife I uncovered the cut of hay we had left last time. If we needed to make a fresh cut in the roof of the stack I stripped the thatch off. For this was no untidy heap of the new-fangled bales, it was a properly-built rick of loose hay with the central layers higher than the outside to keep out the wet, and the steeply-ridged roof properly thatched to protect it from winter's storms.

Fred brought the blade of the hayknife up to razor-sharpness, stroking it carefully with the rubber at a fine angle. His final test was to wipe a single strand of hay gently across the blade. It cut cleanly and he was satisfied.

"Take yer time, Boy, an' get a good sharp on un. You can fair bust a gut tryin' ter cut wi' a dull knife – an' it do take longer in th' end."

The hayknife plunged down into the compacted hay of the stack with a crisp, rasping sound. Fred cut out squares about three feet across and nine inches deep, and I impaled them with a prong and pitched them onto the cart. By keeping the blocks intact and building the load properly we got about half a ton of hay on the small, two-wheeler. A cow eats about twenty pounds each day so our load would last Ernie's seventeen Shorthorns for three days.

Once the load was on, towering over the mare in the shafts, the rope was thrown over and tightened down. The two prongs were slid handle-first in between the sideboard of the cart and the load, with the sharp tines pointing to the rear. They were never stuck in the top or the back of the load – that way they get knocked out and lost on the way home.

Sometimes I'd ride on the load on the return journey, lying flat to avoid low branches with my face buried in the sweet-smelling hay. Flowers pressed flat by the pressure in the stack and perfectly preserved in the drying process looked as if all they needed to bring them back to life was one dewdrop. The scent made me think wistfully of the sunny days of summer when the grass had been cut down by the chattering mower.

Kitty was easy to get back to her warm stable and stepped out on the way back to the yard. She didn't need leading and Fred still trudged behind the cart, his cap pulled down over his eyes and his short, blackened clay pipe jutting out from under his moustache. I often pulled his leg about his pipes – a briar for Sundays and a broken-stemmed clay the rest of the week.

"Clays allus breaks in the stem. 'Sides – I likes a short pipe. Keeps me bliddy nose warm, it do."

By the time Kitty reached the main road by the farm she'd gained fifty yards on the old man. She knew she wasn't allowed to cross on her own and halted at the edge of the tarmac, throwing her head up and down impatiently and cocking her ears back to listen to the heavy tread of her driver's hobnailed boots as he caught up.

Fred went to her head and peered short-sightedly up and down the road. Then, "Come orn then, y'owd booger," and he led her across into the yard. He never addressed his horses by their names – it was always 'y'owd booger' but he wasn't cussing them, it was a term of endearment.

Ernie came and examined the hay critically as Fred and I pitched it off and stacked it under the cartshed. All cowmen have a deep suspicion that there is a plot afoot to supply their milk factory with inferior raw materials and Ernie was no exception. He thought the milk cheque was the only source of income, and that anyone mucking about on the farm with pigs, corn or poultry was wasting time and space and being heavily subsidised by 'his' cows. The hay we brought in always passed muster but there would be some comment and approval, when it was given, was given grudgingly.

"I s'pose that'll 'ave ter do if it's the best you can get me. But what 'appened to all that nice stuff we made off the Fourteen Acre? I han't seen none o' that yet." He turned to me and nodded knowingly. "I knows what's goin' on, Boy, I weren't born yesterday. That old booger'll 'ave the best o' the 'ay saved an' put by fer them bloody old 'osses of 'is, I bet yer."

Fred nodded and winked at me equally knowingly.

"Ar, us knows what us knows, don't us, Boy," he said, and left it at that. It was if he was the master-mind behind the biggest hay swindle in the world. Of course he always made sure that his tallet – the loft over the stable with apertures directly over the horses' hayracks – was full of the best matured hay, but what horseman worth his salt wouldn't? Horses are more particular about their hay than cattle, and the amount they ate was insignificant compared to the consumption of the dairy herd. And it was much cheaper than pouring fuel into a tractor.

By then it was nearly twelve o'clock and time to "put th' 'oss up", give her a small feed of crushed oats and some water and go home for our own dinner. But before I could leave I had to go through the ritual of getting Fred's approval for knocking off. It never varied:

"What's the time, Fred?" I'd ask casually.

"Dunno, Boy. Han't looked." Then he'd pause for a moment – "But I'll 'ave a look now."

He'd haul on the long leather bootlace that served him for a watch-chain, produce a pocket watch (a 5/-d Ingersoll with the chrome rubbed off and the brass showing through), squint closely at its dial and then say as if surprised: "Cor booger! 'tis bliddy near grub time!" Then off we would go, the formalities having been observed.

At one o'clock Fred and I were back in the yard. If there was more fodder needed, or a load of mangolds or straw, we went off to get it. And every three days or so the water cart had to be filled from the village stream. We had no piped water on the farm except in the yard, so water had to be hauled to outlying stock. It was a cold job – difficult not to get gloves and clothing soaked as we dipped water up in buckets to fill the round tank of the water cart. I used to wish we could fill it from the hosepipe in the yard, but that wasn't allowed. Water from the tap cost 2/6d a thousand gallons, but out of the Tillingbourne it was free.

The full water cart sloshed and gurgled its way up Beggar's Lane to the dry and sheltered field between Piney Copse and the railway embankment. This was where we wintered the heifers. The cart was backed up to the old bath we used as a water trough and Kitty was taken out of the shafts. The two props hung under the shafts were let down, crossed over to prevent the shafts slipping sideways, so that the cart was left level.

I pulled the handle on the outlet valve and filled the bath, shutting it off as it came to the brim. The heifers clustered round us inquisitively, interested in the break in their dull routine. Fred stood looking at them for a while, sizing them up and noting their condition. He knew them all individually and speculated out loud on their state of health and how he thought they would milk compared to their mothers.

"That's a pretty little 'eifer, Boy, an' 'er's in good nick too, fer the time o' year. Old Jill's calf, she be. I reckon she'll milk well an' all, when the time comes. Long as she don't turn out ter be a bliddy kicker like 'er old sod of a mother be."

Jill and Dolly were the two nice-looking Shorthorns the boss had bought at the farm sale when old Mr Wonham gave up at Southbrook Farm. The two cows were so alike in appearance that they might have been twins, but in temperament they were totally different. Dolly was quiet and amenable and anyone could handle and milk her, and Jill appeared to be just as docile when you first started milking her. She lulled you into a false sense of security until you had half milked her out, then – WOP! There you were, lying dazed the other side of the cowstall, watching the cats lap at spilt milk as it ran down the gutter.

Ernie tried all sorts of dodges with Jill, but she was lightning-fast and always dead on target – she really meant it. So in the end he had to resort to tying her back legs together with a rope. She objected strongly to this for a day or two but soon gave in. She'd stand quite still provided the rope was loosely round her legs. But if you forgot to put it on...

When Fred had finished lecturing me on the finer points of old Jill's daughter, he pushed through the animals and pointed out his favourite.

"Ar, but you take a look at this one 'ere, Boy. Lovely little roan she be. Pick o' the bunch, she is. You wants a good milker, Boy, you get a nice roan Short'orn."

The old man didn't have much time for the Friesians that were spreading into dairy herds all over the country.

"Leggy bliddy old things, they be. Mind you, I'm not sayin' as 'ow they don't give a lot o' milk, Boy, but fer a nice cow you give I a good old roan. Them black an' whites be such gurt, big devils – 'alf as big again as a Short'orn. Got to cost 'alf as much again ter keep 'em, han't it, Boy? Stands ter reason, dunnit?"

And so on, while we walked over to the hay rick in the corner of the field and threw the day's ration of hay over the fence for the heifers. A constant stream of information and farming lore – heavily biased and antiquated, most of it, but blended with a keen awareness of modern trends. Valuable stuff – the result of sixty years experience.

The heifers' field was bordered on one side by the railway, and at about half-past three a passenger train, drawn by one of the old Southern Railways 2-6-0 'Schools Class' engines, puffed smoothly past up the incline from Gomshall Station and away to Dorking. White smoke drifted high over our heads and Fred hauled out his watch and nodded approvingly because the railway's time agreed with his.

"Come on, Boy, best get back. 'Tis very near pig time."

One of Fred's many responsibilities was feeding the pigs we kept for fattening. He cooked up swill collected from the Abinger Arms – the pub on the corner near the Hammer Clock – in a big copper in the yard, and feeding time was always 'Pig Time'. If the boss asked Fred to do anything different at the end of the afternoon, he'd look at him reprovingly and say to him, "Right-ho, Boss. But it be pig time fust."

Back down the lane past Piney Copse we went, Fred leading the mare and the chains and breechings swinging and jingling in time with the clop of hooves and crunch of boots.

Piney Copse was always a favourite spot of mine. It is a wood of tall trees and dense undergrowth that covers about four acres. I used to play there as a small boy – it was marvellous cover for an Indian brave – but as the country was at war, I was usually an intrepid British commando, hunting down treacherous German spies and Fifth Columnists. Sometimes I'd hear the air raid sirens followed by the drone of aircraft overhead, and I'd be terrified that my game would turn to reality and there'd be burly German parachutists – disguised as nuns, of course – dropping down through the trees around me.

These warlike games and fantasies were mostly inspired by my mother who had worked as a nurse through the horrors of the Great War. She had a fervent patriotic hatred of Hitler's Third Reich and read all the posters warning us that 'Careless Talk Costs Lives'. She saw secret agents all around her and was particularly suspicious of one poor lady in the village who not only had a German-sounding name – Metzger – but also owned two small dachshunds called Ham and Schmidt. What further proof of Teutonic villainy do you need?

The games came back to me as Fred and I walked past the wood, and I pictured myself once more gliding through the trees, my Sten gun ready to annihilate the enemy. 'Ack-Ack-Ack-Ack-Ack!'

"What yer say, Boy?"

"Er – nothing, Fred. I was thinking of something."

"Sounded like 'Ack-Ack' or summat."

"No really, Fred, I was just – um – clearing my throat." I blushed red and scowled, furious that I'd allowed Fred to glimpse my lapse into childishness. I *must* remember that I'm a dignified and grown-up sixteen-year-old farmworker and not a schoolboy. I hurriedly started a solemn conversation about weighty farm matters which lasted until we got back to the yard.

I helped Fred unharness the mare and then fetched a couple of buckets of water for her while Fred climbed up the ladder into the tallet and pushed armfuls of meadow hay down into the horses' hayracks. I waited until Kitty had drunk her fill and then took the buckets away as they shouldn't be left in the stalls. The horses play with them, knock them over and soak the bedding.

Like her master, Kitty was a creature of rituals and habit. She had a little game she played every evening while I watered her. I don't think she ever did it with Fred, but she knew I was a beginner and liked teasing me. She'd half empty the bucket, then raise her head and stare at me, flapping her wet lips and whickering softly through her nose. "Finished, old girl?" I'd say, then put my hand down to pick up the bucket. Directly I did this she'd drop her nose back down and carry on drinking until the bucket was empty, her amused brown eye saying quite plainly, "Had you that time, didn't I?"

"Come on, where you been? Took long enough waterin' them heifers, didn't it? All right are they?"

Ernie was bustling about rinsing buckets after milking. I told him the heifers were all present and correct and ignored the suggestion that Fred and I had been slacking. Ernie secretly envied the variations that Fred got in his work, but he wouldn't have changed places with him.

This inter-departmental rivalry and good-humoured but heavy-handed sarcasm was the basis of good relations between men working on a farm. Each was an expert in his own branch, but with considerable experience in other lines. A ploughman working on arable ground would keep a knowledgeable eye on animals grazing the neighbouring grass and take immediate 'First Aid' action if anything was wrong with them. Then he'd report to the stockman responsible for the animal, *never* the boss. That would have been unforgiveable – akin to 'sneaking' at school. "Taint jonnick," Ernie and Fred would say, meaning anything underhand or despicable.

The proper procedure was as follows:

"Oy Ernie! I see old Polly actin' a bit queer, standin' apart from t'others – so I brought 'er down the yard. Staggery, she is. You wants ter look a'ter 'em a bit better."

"What you mean, staggery? You been interferin' again, you old booger? Let's 'ave a look at 'er."

Then Ernie'd stride into the stall, acting as if the whole business was a waste of time, engineered just to annoy him. But when he saw the cow he'd drop the pose and set about putting things right.

"See the way she's lyin' there a-gruntin', wi' 'er 'ead round 'er flank? Got milk fever, she 'as. Boy, go in the dairy an' get a bucket o' 'ot water, a bottle o' calcium an' the flutter-valve. Fred, 'elp me get some straw under 'er an' get 'er lyin' a bit comfortable."

And after the calcium solution had dripped through the needle stuck in the loose skin on the side of the cow's neck, had entered her bloodstream and made her quickly recover from the potentially fatal calcium deficiency, Ernie would tell Fred to go and poke his nose in elsewhere. But when he told the boss about it, Ernie would make sure that the boss knew it was Fred's prompt action that had saved the cow's life. That was the 'jonnick' thing to do.

Similarly, a cowman working late after milking to help the harvest gang get some corn in during catchy weather, would be teased about getting in the way and not knowing what real work was. But everyone knew that had he not joined in and made the effort, the corn might have rotted in the stook.

I was the only one who got all the kicks and none of the ha'pence, or that was how it seemed. They might have said complimentary things about me to the boss – presumably they did or I wouldn't have lasted long – but I never knew about it. This was right and proper. There's nothing more dangerous on a farm than someone who thinks he knows it all. He's a menace to men, machines, animals and crops – which is why Ernie put me down about everything, including the length of time it had taken to fill the water cart and feed the heifers.

While Fred was feeding the pigs, Ernie and I finished up in the dairy and fed the calves. Then the three of us went into the stall and 'fed round'. This was a longish job as each cow had a bushel skip of Ernie's special mix: rolled oats, ground barley, chopped roots, linseed cake and bran – the whole lot laced with a couple of measures of minerals and a slurp of cod-liver oil before being well mixed with a shovel on the cake-house floor. Then, when this entrée was eaten, the main course – a big pitch of hay – went into the glazed manger in front of each animal – enough to last the night. After that the wheelbarrow was brought in and all the dung was removed from the gutter. The standings were bedded-up with clean straw and the job was done. A final look round; a moment when we paused to listen to the contented crunching noise, and the cowstall doors were shut. They would open again when the boss made his bedtime tour of inspection but for now we said, "That's it 'til morning."

We stood and talked in the dairy while Ernie filled our milk cans. There was the comfortable feeling of a good day's work accomplished, and that tomorrow would be just as satisfying. Then, "Don't be late in the mornin', Boy, see yer termorrer."

"Goodnight, Ernie. Goodnight, Fred," then home to high tea, a roaring fire and 'Take It From Here' on the wireless. Perhaps February wasn't so bad after all.

3

March

If February seems interminable, March is the time of greatest stress on the farm. Everybody – human and animal – is heartily sick of the drab dullness of winter and is longing to be off on the grand adventure of spring. But the cold and the east wind hang on to the last, determined to inflict as much discomfort as they can before withdrawing to their wintry lairs. Animals are fractious and hungry and men get sullen and restive.

I had only been working full-time for three months, and as I was by far the most junior member of staff it seemed to me that everyone's frustrations and irritations were directed solely at me. I couldn't do a thing right, and this, added to my own feelings of discontent, made me think I was the only one at fault. I wondered if I'd made a mistake in taking up farming and wouldn't be better off in a nice warm office, or being a salesman or something. I didn't realise that it was a passing phase everyone was going through, and that it would vanish like magic when the warm weather eventually came.

The only animals little affected by the constant cold were the fattening pigs which Fred tended. They lived indoors all their short lives protected by ample bedding. They probably grew a little fatter during hard weather as they'd burrow into the warmest corner of their yard, only emerging at feeding times. If you peeped into the pig yard when it was cold it appeared deserted. All that could be seen was a large heap of straw from which came much heavy breathing and muffled snoring. But rattle a bucket and the heap would erupt violently as a dozen or so clean, pink porkers raced each other to be first at the trough.

Fred's comments about the weather were the same every morning. "Morn', Boy. Tidn' very bliddy 'ot". And in a cold wind he'd struggle into *two* old army greatcoats. That was his measure of extreme cold –"Tis a two-coat day today, Boy." He had expressions to describe the whole temperature range, through, "tidn' quite so bliddy cold," and "warmish fer the time o' year," right up to the extreme of blistering heat which was, "damme if I ever knowed it so 'ot. Do wi'out me bliddy weskit if it do go on like it."

Cowmen always worry more than other people and Ernie went about his work scowling and muttering to himself. "Knew we should've made more 'ay. Tisn't going to last out. Same every damn year it is. Told 'em, I did, last 'aymakin', but they wouldn't bloody listen."

Then the muttering built up into the nearest we ever got to open rebellion, and he confided in me that he had had enough and was going to leave and get a job where the cows 'was fed proper'.

The tension increased to breaking point when the boss went round surveying the stumps of ricks and clamps we had left, and then told Ernie to cut down a bit, "just to be on the safe side, Ern." This made him furious and when the boss was safely out of earshot he raged at me:

"There you are, Boy, what did I tell you? 'E's done it this time, 'e 'as. When I gets 'ome this evenin' I'll tell the Missus, I will – 'Get packed up, gel,' I'll say. I've 'ad good offers I 'ave, an' I shall bliddy well take 'em. I've been a cowman twenty year, I 'ave, an' I aint never starved no cows yet. I aint startin' now an' that's flat."

"But, Ernie…" I dared to open my mouth and was shouted down instantly.

"Don't you bliddy start, Boy. You don't know nothin', you don't. I should tell 'im straight, next Friday I will, 'Stuff your job', I shall tell 'im. That'll show 'im, that will." And he stamped off angrily and forked extra large pitches of the best hay to the cows to show that there was considerable ill-feeling.

Next morning I crept about my tasks as unobtrusively and efficiently as possible, frightened of re-awakening the storm. And after a while Ernie said, "What's up wi' you, Boy? Got the belly-ache, 'ave yer?"

"Are you really going to leave, Ernie?" I asked timidly.

He stared at me in amazement. "Gorblimey, Boy, whatever give you that idea?" he said. "I give up. I can't make you young lads out these days, I can't. Got no sense o' responsibility, you 'aven't. A good cowman don't just walk out on a herd o' cows at this time o' year, 'e don't. An' I wouldn't want ter let old Bob down, neither. 'E's a goodish boss, 'e is – bit mean wi' the grub sometimes, 'e is, but a bliddy sight better'n most."

"But, Ernie, yesterday you said…"

He interrupted me impatiently. "There's times when I despairs o' ever gettin' any sense into that thick 'ead o' yours. You gets these silly, daft ideas don't yer? Why, I wouldn't leave 'ere, could I? Who'd look after them old cows? An' who'd be 'ere ter straighten you out when you needs it?"

Ernie paused to light the cigarette he'd been rolling and went on: "You know your trouble, Boy? You never listens to them as knows an' you talks too much. Fed up wi' listenin' ter you moanin' day after day, I am. Pack it in an get on wi' yer work."

I stayed meekly silent, resolved to try and do better in future, while Ernie went off shaking his head sorrowfully at the stupidity of modern youth.

Little blow-ups and tantrums occurred from time to time and like fireworks they assumed gigantic proportions and then fizzled out. The boss ignored them. He had to because he often indulged in rages of his own and these were spectacular to watch. One of them I remember in particular. It was caused by the boss's incurable habit of trying to save time and cut corners, which made Fred remark about three times a week, "'E do rush an' tear-arse about so – an' we do end up in a right old pickle at the end of it."

The boss and I were taking a heavy load of dung up to the field with the Fordson. He decided to drive up the muddy side of the field because it was the shortest route. We'd gone about fifty yards when the trailer wheels sank in and the tractor came to a stop with its back wheels spinning in the mud. Instead of unhitching the trailer, pulling the tractor forward on to dry ground and then towing the trailer out with a chain, the boss tried to drive it out by shunting back as far as he could go and then taking a run at the wet patch. He tried it time and again, slamming the gear lever alternately into first and reverse with the engine roaring at full throttle – but the tractor refused to climb out of the deepening rut. Eventually the gearstick gave up the unequal struggle and snapped off short – leaving a four-inch stump sticking out of the side of the gearbox.

The air turned blue, and I was despatched at top speed to fetch a piece of pipe to slip over the stump and jury-rig the gearstick. When I returned, breathless, the boss was scrabbling furiously in the mud, trying to pull out the drawbar pin which was by then deep down and had got bent during all the toing and fro-ing.

"Can't you get the ...?" I started, but he turned on me with a snarl.

"What do you think I'm trying to do? The bloody thing's bent nearly double!" He bent down to the pin again, heaving and straining at it until his face went purple. I retreated to the other side of the tractor knowing that when the pin did come out he'd throw it. I didn't want it to come in my direction.

The pin came out suddenly and the boss flew over backwards, ending up sitting in the liquid mud. He stared at the drawbar pin in his hand for a moment and then his reason snapped. "Bloody bastard pin!" he shouted, and it went sailing high over the hedge into the next field.

It didn't take long to drive the tractor forward, put the chain on and drag the trailer out, but when it came to coupling up again we had a problem.

"Come on, hurry up and get the pin in. We've wasted enough time as it is," he snapped, and I explained as gently as I could that he's thrown our only drawbar pin over the hedge.

It took twenty minutes of searching to find the missing pin, and while we were doing it I kept very quiet indeed.

The best thing about the boss was that his rages died down as quickly as they had flared up. If he was mad at you about something he'd come right out and tell you about it in no uncertain terms. Then when the message had been received and understood, we'd all be friends again. Much better than moping and nagging for days on end which makes everybody miserable. And it was accepted that everyone needs to let off steam now and again in times of stress. Everyone but me, of course. I was only 'Boy' and hadn't yet earned that right.

The small outbursts always concerned animals, crops or machines. They were never about wages, working conditions or the rights of the workers. And if any union-minded twit had tried to divide us up into 'management' and 'shop-floor', he'd have been booted out of the yard a bit quick. We were a team and although the boss paid our wages (Ernie £7, Fred £5, me 30/- a week), we worked for the farm and not for him. He also worked for the farm, much harder than anyone else, and we respected him for it.

Our attitude to the Farmworkers' Union was strongly influenced by the fact that the only two men in our area who were members were considered to be useless troublemakers. The boss commented on this one day and said that if anyone working for him joined the union he'd sack him. Being a bit 'Leftist' in those days, I objected to this and said, "That's not fair – you're a union man yourself."

"I'm bloody well not," he replied indignantly, "I've always voted Tory."

"You're in the NFU, aren't you?" I said, greatly daring, and he looked at me consideringly.

"I see what you mean," he said, "but look at the difference between Fred and Ernie who aren't union men and that useless bugger up the road who is. Do you think I'd want to have anyone like him on the place? Why, nobody would want to work with him for a start."

I couldn't answer that, it was true.

We didn't run out of fodder and, in spite of Ernie's gloomy predictions, none of the cows died of starvation. I thought that we'd been extremely lucky and I didn't realise that it wasn't good luck that had kept us going, but good planning by the boss. And soon, with one of those dramatic changes so characteristic of English weather, the wind swung round to the south and spring started to appear.

The change was noticeable all over the farm.

Milk yields went up because the cows were getting a picking in the meadows and the cowman strutted about with a self-satisfied grin.

"'Nother couple o' weeks an' there'll be a bite about, Boy." he said happily. "They'll soon be able to get their old tongues round it. Good job I cut down the hay when I did or we might've run short. But when you been in the job as long as I 'as, Boy, you knows 'ow ter go on".

All of a sudden there was a lot of field work to be done. I revelled in this because I was sent out on the old Fordson, chain-harrowing the grassland to shake up the roots of the grass, pull out moss and spread the dung-pats; working the arable ground down to a tilth for corn sowing; and deep-cultivating the root ground to mix in last autumn's ploughed in dung.

The boss and I were the only ones to drive the tractor. Ernie was far too busy with the cattle, and Fred had never learnt to drive.

"I tried one once a few year ago afore the war, Boy," he told me, "but the bliddy old thing bolted into the 'edge wi' me an' I give it up there an' then. An' they aint no value any'ow – you wants real good work you got ter use a pair o' 'osses."

But no matter what Fred said, the 1931 Fordson with its iron front wheels, enormous thirst for paraffin and unpredictable steering (there was half a turn of play on the steering wheel), seemed to me the essence of modern farming. I put up with its idiosyncrasies, sudden breakdowns, sulky reluctance to start from cold, and total refusal to fire when hot or over-choked, as I didn't know any better. I thought that all tractors were like it. I assumed it was normal to stop every half-hour to clean oiled-up plugs and topping-up with a gallon or so of engine oil each day was routine. There was no instruction book so I didn't know you were supposed to change the engine oil in a tractor – not that it mattered much, the engine used so much oil that it changed itself about twice each week.

I felt like a real farmworker when I was on that Fordson. There was I – for some of the time at least – in control of twenty-five horsepower. The beat of the slow-revving, side-valve engine and the distinctive smell of burning paraffin were intoxicating. The feeling of power was increased by the fact that you had to achieve and maintain mastery over the beast – it wasn't like a modern tractor controlled by pressing buttons and with a languid hand on the power steering – it was a fight to the finish between man and machine. If, at the end of the day you were exhausted by the conflict, the satisfaction of having won made up for any aching muscles.

When the ground had warmed and dried sufficiently we started drilling spring corn. But we didn't just rush out and begin – there were a lot of formalities to be gone through first and most of these centred around old Fred. He was always consulted when important field operations were being discussed, and with good reason. Not only did he have more years of experience behind him than the rest of us put together, but he had been on the farm for years before the war and knew every field better than most people know their gardens.

The boss brought the tea down early that morning so Ernie and I knew something was up.

"Can you manage on your own today, Ernie? That top field's just about ready and it'll take the three of us to drill it and harrow it in."

"You go ahead, Boss," said Ernie. "Just let 'im 'elp me get them churns out for the lorry, then you can 'ave 'im an' welcome. Get on a sight quicker, I will, 'thout 'im under me feet." But he winked at me as he spoke so I knew he was pulling my leg.

"Drilling!" I said excitedly. "Shall I go and tell Fred?"

"No." said the boss firmly. "You won't *tell* Fred anything. I'll *ask* him myself."

Just then Fred came into the cowstall for his tea. He'd finished seeing to the horses and pigs earlier than usual and I wondered how he'd known the boss wanted to go drilling.

"Morn', Boss. Morn', Ern. Morn', Boy." Fred took his cup from the tray and stood there supping noisily through his moustache. He eyed us speculatively but said nothing. If he knew something was up he wasn't going to let on, that wasn't his way.

"Fred," said the boss, "think that top field under the Downs'll be fit to go today?"

Fred spat some tobacco juice, looked doubtfully out of the cowstall door at the sky and replied, "It might well be, but us best goo an' 'ave a look fust, 'adn't us?"

"I looked at the ground last night, Fred," said the boss, "and it seemed to me it's ready."

"Ar, it might be an' all," Fred sounded dubious. "but that were last night. That's funny old ground up there on that chalk. Might be different altogether, 'smornin', it might. An' you asked I fer my 'pinion on it – I aint seen the ground meself so I aint goin' ter say one way or t'other."

"Well – let's go and have a look now, Fred," said the boss diplomatically, "and then you can tell me what you think." He led the way out of the cowstall.

I desperately wanted to go with them and looked pleadingly at Ernie. He relented.

"Go on then. I 'spect they got ter make some big

decisions an' you'd best be with 'em ter put 'em right. But, hey!" – I skidded to a stop in the doorway and looked back – "You make sure you're back 'ere afore eight o'clock. We still got to put them churns on the stand, drillin' or no bloody drillin'."

I caught up with the boss and Fred and we walked up the lane together. Fred was telling the boss all about farmers he'd known who'd drilled corn before the ground was properly fit, and how massive the crop failures had been as a result. I piped up and said that I'd seen someone drilling the previous autumn and the land had been wet and very cloddy. Would that crop fail? I asked.

"That's different altogether, Boy." said Fred. "You puddles winter wheat in on a real rough old tilth. It don't mind the wet, an' the clods do 'tect it from the frost. It got all winter to get 'stablished, an' when the ground do warm up in spring you gives it a roll an' it do go away like ninepence. Spring corn's different. You got ter get the land right 'cause it's got ter grow on proper straight off."

We walked up Beggar's Lane, through the railway arch and passed the four-acre 'Little Field' without comment. The four acres of heavy clay was still lying cold and heavy in the furrow and would take another three weeks before it, too, felt the awakening of spring and demanded our attentions. Next up the lane was the Horse Meadow, six acres of permanent grass with a dewpond in the corner which never dried up in the hottest summer. That field hadn't been ploughed for a hundred years, but the ridges and furrows left by the ploughmen of four generations ago were still plainly visible. Being mid-way between the heavy clay and the chalk of the Downs it was just starting to move and Fred spared it a glance as we passed.

"Be a bite there soon, Boss. Unless you leaves un fer 'ay. Makes a pretty bit o' medder 'ay fer the calves an' 'osses, that do."

"Good grazing too, Fred." said the boss. "Pity it's so far from the farm or we could put the cows in it."

"Oh Ar." agreed Fred. "Fat a bullock in no time at all, that ol' medder will."

But they weren't interested in the Horse Meadow and pressed on to the Top Field which was the object of our particular attention that day. And it was there that I witnessed one of the first rites of spring – one of those moments of pure magic which I was just beginning to understand and appreciate.

The two experts stopped in the gateway and looked thoughtfully up and down the field. Then they set off slowly, looking at the ground all the time, to separate ends. They paused and looked around, then walked slowly back, still looking at the ground, to meet in the middle of the field. Still without a word they crouched down and each placed a hand flat on the moist earth to feel for warmth. They stared vacantly at each other for a moment or two, their eyes out of focus, then got to their feet, brushing the dirt from their palms.

Fred nodded. "That'll go nicely, Boss." Then he added, "If it don't rain, like." He always hedged his bets.

I asked them what they had been looking for in the ground but they couldn't tell me exactly.

"'Tidn't just a matter o' bein' dry enough, Boy, 'tis summat else. You just knows when it be right. You'll learn, you will, after a year or two."

I now know what Fred meant. The two men weren't just testing the soil temperature and examining the structure of the soil – their eyes looked at the ground but they didn't really see it – they were dowsers, divining the field without a hazel twig. A field is a living thing and has a spirit and being all of its own. The soil teems with countless millions of tiny living organisms, each one excited and stimulated by the coming of spring; each one galvanized into furious activity and sending out an infinitely weak signal and message. When all the living bodies send together on the same wavelength that faintest whisper becomes almost a shout in the mind of the countryman. The tiredness of winter drops away – he knows that the message has come through and he gladly responds. Why or how he cannot tell you. He just "knows when it be right."

Fred started putting on the 'bag-muck' with the horse-drawn spreader while the boss and I got the drill ready. This was an ancient 'Massey-Harris' machine converted for the tractor with a drawbar instead of the original two-horse pole. It was narrow, slow, had a small seedbox and kept falling to pieces. It worked, after a fashion, and we drilled about six acres a day with it which was the average field size on that small farm.

I wasn't allowed to drive when we were drilling as I might have left misses which would have been obvious to the neighbours for the whole season. But I was entrusted with the job of harrowing-in after the drill, having been given a short lecture on how to 'goo on about it.'

"Don't goo too fast, Boy, or them 'arrers'll jump – an' not too slow neither or you wunt get done. Don't leave no gaps atween your passes, an' don't overlap too much or you'll pack the ground. If your 'arrers blocks wi' trash you make sure you stops an' clears 'em or the lumps o' trash'll dig in an' rake that seed clear out o' the ground – then they bliddy rooks'll 'ave it. An' when you finishes round the 'eadland you drive right to the gate afore you un'itches the 'arrers. Leave a wheelmark an' the boss'll goo mad at yer."

When the field was drilled there were some anxious days when the boss patrolled with his gun. Rooks descended in clouds on the newly-disturbed earth and probed with their long bills, digging up the succulent sprouting grains. We used to shoot some and hang them up on long sticks – about one to every acre of ground – where they dangled and swung forlornly in the wind, a dreadful warning to the others that crime does not pay. Rooks are highly intelligent and took heed, but the same could not be said for rabbits, which were the next pest to take over the attack on the crop directly the first green shoot appeared. You could hang up as many dead rabbits as you like and the rest would still munch unconcerned at the growing crop. There were thousands about in the days before myxomatosis and no matter how many were killed on the farm, they still came in droves off the close-cropped turf of the Downs. It was a case of shooting as many as you could, maintaining a constant presence to keep them on the move, and praying for warm weather so that the crops grew away from them.

In the old days when labour was cheap, farmers used to employ young boys who were stationed in the fields throughout the growing season to scare pests away from vulnerable crops. They were there from dawn to dusk and had to constantly shout and holler and wave wooden rattles which clatter loudly when you spin them round – probably the origins of the saying 'Going like the clappers'.

Fred said that was how he had started on a farm long ago in the reign of Queen Victoria.

"Bird starvin' I was, Boy, an' I got paid a shillun an' a tanner a bliddy week fer it."

I thought how lucky I was to have to have been born in an age when beginners were paid a real wage for a week's work. I got thirty shillings a week – twenty times the amount Fred got when he started – and the week was shorter at only fifty hours. I gave my mother a pound for my keep and had ten shillings pocket money. But I grumbled to Fred that my cigarette tobacco was expensive at 2/7d an ounce and he agreed with me.

"Oh, Ar – things is terrible dear now, Boy. Why when I were a young carter an' just got wed, us got by on ten shillun a week – what you spends on yer fags now – but us lived well enough. Mind you, we allus 'ad a pig in the sty an' killed un in th'autumn, like, an' a good bit o' 'llotment fer greens an' spuds an' that."

He paused, smiled reminiscently and then went on:

"O' course, bacca were on'y tuppence a' ounce an' beer were tuppence a pint. Cor booger, Boy, you could git pissed as a lord fer 'alf a crown!

4

April

The cowman and I started milking a little earlier in the mornings during the hectic month of April. We got on quicker, too, as I was becoming more proficient and could take a full share of the milking instead of being limited to the easy-milking, quiet old cows. Ernie still insisted on doing the fresh-calved 'heavies' himself. The beginning of the lactation is the time when experience is needed to distinguish between 'nature', the normal stiffness and distention of the calver's udder tissues, and the abnormal inflammation and swelling of mastitis.

An expert cowman, Ernie could detect an illness in a cow before any symptoms were apparent that a vet would recognise. He had the 'stockman's eye' that can only be acquired by long experience. I once saw him staring fixedly at a cow that was munching happily at her hay in the cowstall and asked him what he was looking at.

"Dunno, Boy," he said, "but there's summat amiss wi' 'er. Reckon she'll be off-colour tomorrer. We'll 'ave ter watch 'er."

Fred was pleased that I was doing more in the stall as it meant that he didn't have to help with the milking. He was getting too old for that game, he said. "Best leave it to you young uns. I done my bliddy share." And he was busy in the mornings getting ready for the boss to come down to the yard bubbling over with enthusiastic plans for the new day.

The morning tea session turned into a planning discussion between the two arable experts, and when the day's work had been arranged the boss rushed back to the farmhouse to gulp down his breakfast before sallying forth to put the plans into action.

It took two of us to lift the two-hundredweight sacks of seed and fertilizer onto the iron-wheeled, flat-bed, turntable wagon called 'the trolley' and I was usually sent to help Fred load. Ernie had a little grumble about having to milk the last few cows on his own, but he realised the importance of getting the drilling done while the weather held.

"Shall I ask him if I can stay and help you finish?" I asked, hovering uncertainly in the doorway.

"No, Boy." was the reply. "Do like 'e said. I s'pose I can manage wi'out your valuable 'elp – an' anyway, I can see you be a-bustin' ter be out there."

That was it. He knew I was longing to get away from the routine of the cowstall and out into the fields where so many exciting things were happening.

The ground had been given its final harrowing, either with the tractor or with the horses nodding their heads across the field in the old, proud way – Fred stumbling over the clods behind – and we got busy with the drill.

At last we'd got a new combine-drill. The boss was tired of mending our old Massey-Harris, horse-drawn, seed-only machine. He had converted it to the tractor with a drawbar instead of the horse pole, and the headlong, four miles an hour of the Fordson was too much for it. It kept falling to pieces and holding us up, wasting valuable drilling time. So after a cautious visit to his bank manager, the boss ordered a brand-new 'Sunshine' combine-drill.

"It's the latest thing, Fred." he said proudly. "It puts the seed and fertilizer on all in one go. And as the fertilizer goes right alongside the seed, the young roots get a good start."

Fred wasn't impressed. "I don't know so much, Boss. When you spreads yer fert'lizer in the seedbed, th' old way, them roots've got ter spread out ter look for it, like, an' they gets big an' strong doin' it."

"Mind you," he added, "I knows our old 'oss drill aint no cop no more, but them new ones is made to go be'ind tractors. Goes too bliddy fast, they does. You can't do real good work like that."

"Silly old fool." said the boss. "We've got to keep up with the times." But he waited until Fred was out of earshot. He didn't like to offend the old man.

When the drill was delivered, in all the glory of its new yellow paintwork, we left it out by the road for the neighbours to marvel at for a couple of hours, then took it up the lane to the field. The boss drove the tractor and I rode on the varnished footboard at the back of the drill.

Just as we went carefully through the gateway – the drill was wider than the old Massey-Harris – I heard a tinkling noise and noticed that the shoe of one of the thirteen coulters was dragging loose. We stopped and examined it.

"Look at this!" said the boss. "Blasted nut's come undone. They can't have tightened it up properly at the factory."

We checked the rest of the coulters and found that they were all loose. And more than that – *every* nut and bolt on the machine was slack. The bumpy quarter of a mile up the lane had nearly shaken the drill to pieces. The reason was obvious – there wasn't a single spring-washer anywhere on it.

There was no point in tightening up the loose nuts without washers as they would have come slack again in a few minutes, so I was despatched to scour the village for spring-washers. "Go to the garage, the blacksmiths and the builder. And get a shift on – I want to get some blasted corn in today."

I sped back to the field, panting on my bike up the steep lane, having visited all the likely sources of spring-washers in the village. I had to be particularly careful at the garage where the mechanic was a notorious practical joker. He had a nasty habit of sending gullible youngsters off in search of tins of elbow-grease, left-handed screwdrivers and the like, and any such delay on *that* day would have driven the boss to the verge of apoplexy. My pockets bulged with assorted spring-washers, nuts and bolts and all the spanners I could lay my hands on. I turned them out and we got to work. We took off every bolt on the drill, fitted spring-washers and tightened the nuts up properly. It took us about an hour and a half, and we were just finishing when Fred came down the lane leading two horses. He stopped to have a gloat.

"Whoever made that 'chine by the way?" he asked me.

I read the maker's label on the drawbar. "'Sunshine Sundrill'. Made in Australia." I announced.

"Ar, that'll be it then." said Fred. "They don't understan' mechanicals out there, Boy, it be all sheep an' kang'roos an' rabbits an' that." He turned to leer evilly at the boss who'd just crawled out from under the drill. "I finished 'arrowin'- down wi' the 'osses, Boss, an' I 'spect you'll be a-startin' wi' that time-savin', wunnerful new tractor drill." He spat thoughtfully on the wheel of the drill. "Boogered if that han't saved you a lot o' time today."

The boss didn't reply but his gaze rested with a savage intensity on the back of Fred's shabby black jacket as the old man plodded off down the lane between two horses, chuckling with glee. The jacket didn't quite burst into flames but I swear I saw it beginning to smoulder.

Despite Fred's sneers about it, the new drill made a

considerable difference. The greater width, bigger seedbox and increased working speed enabled us to finish off the spring sowing during the first few days of April.

"No 'Cuckoo-Corn' this year." said the boss with satisfaction. "Now we can get on and get the roots in."

His reference to the cuckoo meant every farmer's dread of still having sowing to do when the first letter appears in 'The Times' – usually from a gentleman of the cloth in Gloucestershire – announcing the arrival of that wily bird. Late-sown crops yield less than timely-sown ones, but if it is a late season and everyone is sowing in April, then the cuckoo will be late too; so it is a point of honour to have the drill put away before cock meadow pipits and reed warblers start eyeing their wives dubiously.

There was none of the rush and scurry of corn drilling about sowing mangolds, fodder beet and kale. Much more attention was paid to getting the ground smooth and flat with a very fine tilth.

"That's got ter be down like a' onion-bed, Boy." said old Fred. "Get a lot o' clods on the top an' you'll 'ave the devil's own job wi' yer singlin' an' flat-'oein'. An' you can't drill straight when it be lumpy, then when you puts yer 'orse-'oe through the crooked lines she'll cut out 'alf the plants."

He made sure the job was 'done proper' by giving the root ground its final rolling with the horse-drawn flat roller. "One or two 'oof marks," he said. "but none o' them gurt ruts what that tractor do leave." Then he got on with drilling the roots, driving the antiquated, three-row root drill with a mare in the shafts. The rows were as straight as a gun barrel.

"Can I have a go, Fred?" I asked, wanting to try my hand at anything.

"Not yet you can't, Boy." was the reply. "You'd make my work about as straight as a dog's 'ind leg, you would, in a couple o' turns." He always talked of 'turns' and 'passes'. A 'pass' is once across the field; a 'turn' is two 'passes' – across the field and back to the same headland.

The new corn was greening-up nicely in the fields under the Downs, and while Fred was busy drilling roots I spent long days harrowing and rolling. The heavy harrows looked as if they were doing terrible damage to the young wheat as they ripped along, but they were, in fact, stimulating it. They parted the little plants, making them 'tiller out', or produce five or six shoots from the one seed that the men said survived from four planted –

One fer the rook, an' one fer the crow:

One ter rot, an' one ter grow.

Also, the harrows weeded the corn, knocking out the seedling weeds that were germinating between the rows. This was the only form of weed control that was possible, apart from careful pre-sowing cultivations. Selective herbicide sprays were in their infancy and had not been generally adopted. And to a certain extent they were not necessary. Weeds were tolerated because their depressing effect upon grain yields was not noticed so much in the lighter crops of those days; any growing in the corn at harvest time being safely carted off the field in sheaves to the rickyard, where their seeds could be bagged-up off the threshing machine and then burnt. The modern combine is the villain of the piece – it cannot deal with weed seeds and blows them back onto the ground where they germinate and 'increase and multiply'. It's the combine that forces today's farmer into the widespread use of expensive chemicals, not his desire to annoy conservationists.

Rolling firmed the soil round young plants, drawing up moisture from beneath, making life difficult for hungry wireworms, and squashing leatherjackets and slugs that were lurking under bits of turf and stones on the surface.

It was lovely driving the Fordson up and down the fields in the spring sunshine. The alternate light and dark stripes left by the roller were pleasing to the eye like a newly-mown lawn, and the tractor ran well. After a winter of idling about doing carting jobs, the recent bout of hard pulling with the drill had blown the soot out of the engine and virtually decoked it. Rolling was easy pulling but just load enough to keep the engine temperature high and stop the plugs from oiling-up too often.

The main hazard, to me, was that the steady crackle of the open exhaust and the grumbling whine of the Fordson's worm-drive differential made me very sleepy. I kept dozing off. It wasn't that the job was boring – far from it – but I couldn't help nodding from time to time. I'd stop when I felt myself getting glassy-eyed and run round the tractor to wake up, but on two occasions I left it too late.

The first was when I sleepily missed the turn at the headland and drove the tractor deep into a thorn bush. A sturdy blackthorn stem pushed the starting-handle back against the dogs on the crank pulley, and the terrible screeching rattle made me jump out of my skin. In my panic the only thing I could think of was to stop the engine and put an end to the frightful noise, so I earthed the magneto. And that, of course, was that. There was no self-starter on a Fordson – they were started by swinging the handle – and the handle was buried in thorns.

I had to walk all the way back to the farm, confess to the boss, and get him to help me cut the thorn bush away with an axe and a saw so that we could swing the handle.

When we came to start the tractor again I made another awful discovery. Fordsons ran on paraffin vaporised by the heat of the exhaust manifold, and had to be started on petrol when cold. The engine had cooled while the tractor was stuck in the thorns and I'd forgotten to top-up the petrol tank that morning after using what was in there. The tank was empty and I had to walk down to the farm again to fetch some petrol, wasting yet more valuable time.

On the way down the lane the boss told me how kind, considerate and thoughtful he was. He had feelings of genuine compassion and sympathy towards forgetful boys who made silly mistakes. Were I working for an ordinary, run-of-the-mill farmer – interested only in profits and efficiency – I would, he said, undoubtedly have been strangled, jumped upon with heavy boots and then instantly dismissed. But he wouldn't do anything like that – he was much too nice a person. But, By God! – Next time…

And the second time I made a fool of myself through getting sleepy it was right in front of the boss. He had sent me off to harrow some root ground down and had warned me about keeping straight. Normally it doesn't matter if you don't keep straight with harrows, but he liked it in fields – like the one I was in – where there was a footpath patrolled by critical neighbours.

Bob Newman was the type of boss who, if he wasn't working alongside us, walked the farm the whole time. He'd appear suddenly through the hedge, not because he snooped about trying to catch people slacking, but because he took short-cuts everywhere and wouldn't waste time walking round to a gateway. On the day I was harrowing I saw him climb over a hedge and stand watching me, at the far end of the field, as I drove towards him.

I lined up the radiator cap of the tractor on the distant figure and concentrated on driving as straight as possible. I'll show him, I said to myself.

When I was about twenty yards from the boss he started shouting and waving his arms about. I couldn't hear what he was saying over the noise of the exhaust but stopped the tractor and sat there mystified. He came up and leant on the mudguard.

"What the thundering bloody hell do you think you're doing?"

I was even more mystified but hazarded a guess.

"Harrowing?"

"Have you looked behind you during the last hour or two?" he snarled.

I looked over my shoulder. There, stretching back across the fields, as straight as if they'd been drawn with a ruler, were the parallel tyre-tracks of the Fordson. They shouldn't have been there – they should have been wiped out by the harrows. But the harrows weren't behind the tractor – they were sitting on the headland at the far end of the field. I'd been so engrossed in my beautiful accurate steering that I hadn't noticed when they came unhitched.

The boss then delivered a short sermon to his captive congregation of one. His theme – and I remember it very clearly indeed – was reminiscent of Luke IX:62 and was that whosoever puts his hand to the plough, the cultivator, the discs, the harrows or any other damn thing hitched behind a tractor, and who *doesn't* look back every few seconds, is not fit for the Kingdom of Newman.

While all these exciting things were going on in the fields the cows still had to be milked twice a day. But the most momentous day in the dairyman's calendar had, at last, arrived – the day when it was decided to turn the herd out at night.

The date was only agreed after much thought and earnest debate. The problem was that the cows, through being kept in the warm shed all winter, had thin coats and couldn't stand the cold. If you turned them out too early, had some late frosts, and had to bring them in again, the milk yields would drop disastrously; and was the grass growing strongly enough to carry the herd without being supplemented with the little bit of hay that was left?

All these factors had to be taken into consideration and it was the boss who decided in the end. The cowman couldn't make up his mind because the welfare of his darlings was at stake; I was too young and inexperienced to have any say in the matter, and Fred had an artful way of expressing an opinion without committing himself one way or another.

"I knows what I'd do if 'twere up to I," he said knowingly, "but then I baint cowman, be I?" And he left it at that. He'd have made a first-class politician, would Fred.

The evening the cows were turned out for the first time was very pleasant as we didn't have all the chores of hand-feeding and bedding-up to do. We just let them out of the stall, pushed them across the Marsh onto the Hangers, shut the gate and we were finished.

Ernie, Fred and I leant on the gate and watched the cows graze. Ernie made gloomy predictions as to how many chills he'd have to treat in the morning, and how much the milk would be down. Then with a curt "Goo'night", he strode off on the long walk past Southbrook Farm to his cottage at Tower Hill in Gomshall.

I was worried about the cows and sought reassurance.

"Will they be all right, Fred? Ernie's worried about them. Have we turned out too early?"

"Don't you worry yer 'ead, Boy," said Fred.

"Old Ern's too bliddy soft wi' them cows. Allus wants ter wrap 'em up in cotton-wool, 'e does. Should've turned out a couple o' weeks back – a bit o' cold snap at night don't do 'em no 'arm."

"I hope you're right, Fred, but Ernie'll moan like hell if anything goes wrong." Fred snorted scornfully. "Don't you take no notice o' Ernie's moanin's, Boy. 'E's a cowman. Cowmen be allus a-moanin'. I knows – I been a cowman myself afore I got sensible an' turned to 'osses."

We walked back to the yard and Fred spoke again.

"Come on, Boy, you can 'elp I bed-up they two 'osses. I aren't goin' ter turn they out fer another week or two yet – 'twouldn't be right."

5

MAY

Walking to work on a May morning is a pleasure that makes youngsters whistle as they go. The whistle joins rather than competes with early birdsong, and the mind seems more alert and perceptive, taking in all the delightful details of the countryside at its most attractive and promising time of year.

The air had a nip in it – almost a frost, but stimulating rather than cold – as I went to get the cows in for milking. Everything was covered thickly in dew from the dense mist rising from the watercress beds beside the road. The mist was always there in the mornings; it filled the whole valley and came from the warm, spring water of the cress beds meeting the cool night air. It rolled away miraculously as the sun came up over the Downs, and lie-abeds never saw it. I had it all to myself as I walked across the Marsh to the Hangers, the steep meadow where the cows grazed at night.

I started calling the cows before I reached the gate, hoping that they would be waiting for me and I wouldn't have to search the field for them. 'Cup, Cup, Cup Aloong! Come Ooorn, Then!' I copied the call Ernie used as it was the sound the cows recognised. Cows are very much creatures of habit and like everything about them to be constant and familiar. Any deviation from their normal, placid existence makes them edgy and irritable – they stand alert and watchful, instinctively fearful that wolves will lope out of the forest and attack.

But there were no wolves that misty morning, just one or two friendly bovine shapes waiting by the gate, while from the hillside above, the sound of plodding, deliberate hooves as the Shorthorns responded to my call.

I opened the gate and watched them walk out across the Marsh in single file, each cow treading carefully in the footsteps of the one in front, their tight bags swinging ponderously, showing the results of a hard night's cudding. They stopped at the stream for a long drink then cropped the lush grass at the water's edge, hooves sinking hock-deep into the black mud, while they waited for me to drive them on over to the yard.

The last cow came through the gate at a run, anxious not to be left behind, but more worried about missing a succulent mouthful or two that the others were enjoying while we waited for her. I closed the gate and ran my eye over them, checking that they were all there. Flossie, Whitefoot, Jill, Dinah, Daisy, Valentine – and so on, right through the seventeen. But, wait a minute, there were only sixteen. A quick check through to make sure I hadn't mis-counted and, sure enough, there was one missing. I ran through the names in the order they stood in the stall to see who it was. "Judy, Nancy, Daisy, Gentle, Folly, Prim..." It was Primrose, the old red Shorthorn with a touch of Guernsey in her breeding. Ernie had said, the night before, that he thought she was near calving. I thought I'd better find her and make sure that she was all right.

The rest of the herd were halfway across to the yard – I could hear Ernie calling them from the gateway – so it was safe to go and look for Primrose.

I climbed the hill, following the line of the fence that divided the two halves of the Hangers. When I got to the massive oak tree at the top I came out of the mist into the clear, spring morning. The cotton-wool of the mist stretched away below me, down the Tillingbourne valley towards Guildford; the hazy, blue hump of St Martha's, above Chilworth, a distant island in the sea of white.

It was difficult not to stop for a while and admire the view, but over by the hedge at the far side of the meadow I could see Primrose, and by her side, a small jerky movement as her calf tottered about trying its new legs. I walked over to them remembering what I had been told.

"Walk up to 'em slow, Boy, an' don't goo too close. Talk to 'em, make sure they're all right an' then leave 'em be. Cows what's just calved is funny devils an' don't you ferget it."

I'd protested that I wasn't frightened of cows and got another of Ernie's short lectures.

"An' I aint frightened o' motor cars, neither, but I don't go jumpin' in front of 'em – I don't take no risks, like. It's the same wi' cows. You shouldn't never be scared of 'em, but you always respects 'em."

He told me that a cow with a new calf could be more dangerous than a bull. But – above all – when dealing with any animal I should never show fear. The fear would be communicated to the animal which would react instinctively and immediately. This lesson was brought home to me when I first started milking. We had a newly-calved heifer that was a bit hasty with her feet. She wasn't a real kicker, just nervous. Ernie always milked her himself, and I had to stand behind her, holding her tail up straight to stop her fidgeting. One morning, when it was dark, I went into the stall and sat down to the heifer by mistake, thinking she was the old, quiet cow I was supposed to milk. She stood like a good-un, and I had nearly finished her when Ernie brought the lamp into the stall. I looked round and saw the distinctive white face of the nervous heifer looking at me. I froze with panic and then started to ease myself away from her. She immediately lashed out and I ended up flat on my back in the gutter. Ernie was unsympathetic.

'If you're daft enough ter milk the wrong bliddy cow, Boy, you should've 'ad enough sense ter stick it an' milk 'er out. Jumpin' up like you did you was askin' for it."

Back at the cowstall Ernie had tied the cows up and started milking.

"Where the 'ell you been? Lazin' about dreamin' again I s'pose. Where's Primrose – calved 'ave she?"

I told him Primrose had a heifer calf and he nodded approvingly, his head up against the flank of the cow he was milking.

"Good old girl, she is. Always 'as a nice calf. Get 'em in after milkin', we will."

When the full churns had been labelled and rolled out to the milkstand ready for the lorry – there were twice as many now that the herd was in full spring production – Ernie and I walked up the hill to where Primrose stood licking her calf. We got behind them and coaxed them down to the yard.

Some new-born calves will be off like a rocket when they first see a human and you have to run fast to catch them. And some cows, too, will revert to the behaviour of their wild ancestors and lead the baby off at full gallop when they think danger threatens. Grazing animals like cattle, horses, deer and antelope are constantly on the move and their best defence against predators is flight, so the young are born with long legs and the ability to run and keep up with mother and the herd. The slow ones end up as a meal for the scavengers that trail a herd of wild cattle, and the speedy survivors grow to maturity and breed strong, long-legged calves of their own. Very nice for Mr Darwin and his theory of natural selection, but not so good for the cowman panting along in his gumboots in pursuit of an errant calf.

The best time to catch a calf when they're like that is when it's still wet, but that has its problems. You get the little animal to its feet and it stands there swaying. Any attempt to get it to take one step towards the gateway results in it collapsing progressively like a house of cards falling, to lie there legs a-straddle, rolling an indignant and reproachful eye at you. Then you have to pick it up and carry it home.

It's easy with two of you. You link hands under the calf's chest and belly and carry it between you, the long legs dangling limp and the little head swinging owlishly from side to side. Mother trots anxiously behind, nuzzling at her baby from time to time – or sometimes she doesn't.

For some reason or other, some cows, when their calf is lifted from the ground, lose sight of it. You have to watch them over your shoulder and the moment they turn away, you put the calf down and stand away from it so that they can see it again. It's infuriating to see a cow cantering back the way you have just laboriously carried her calf. She will always go right back to where the calf was born, thinking it will still be there. It must be vastly amusing to an onlooker, seeing two men standing with a small calf between them; a cow disappearing at speed over the skyline, and the men making frantic calf noises to try and entice the cow back.

"Baaaa! Ber! Baaaa! – Come on old girl!

Ber! Baaaa! Baaaaaaa! – Blast the old devil!"

Then there is nothing for it but to carry the calf back to the cow and start all over again.

Primrose, though, was a steady old cow and her calf was well-behaved. They walked sedately to the cowstall where Primrose was tied up in her standing and the calf was taken away to join the others in the barn.

Although the calf got its mother's milk for the first few days we always parted them as soon as possible. People who don't know animals and wrongly attribute human feelings and emotions to animals, think that it is cruel to do this, but it is, in fact, the kindest and most sensible way. If the calf is left on the cow for two or three days she will get so attached to it that she will go over or through the strongest fence to get back to it when they are eventually parted. If she doesn't injure herself badly in doing so, she will pine and be miserable for days on end. That is real cruelty. And the calf pines as well – having got used to sucking from the cow it will not take kindly to milk in a bucket. But take the calf away within twelve hours of birth and the cow will have forgotten all about it by the following morning; she's glad to settle back into her normal routine.

Ever since Fred finished drilling the roots the boss had been walking the root ground waiting for the first seedlings to show themselves. Fred sharpened up his favourite hoe, a piece cut from an old scythe blade and set in a hoe frame, and told me that he would show me how to do some real work for a change.

"'Ave to bend yer back now, Boy. Still, will do you a power o' good – better'n ridin' around on that old tractor all day."

I wasn't sure that I agreed with him about the tractor, but hoeing couldn't be all that hard, I thought. Anyway, I'd show him that modern youth could compete with crabbed old age any day of the week. I was quite pleased when the boss told me to go with Fred and start hoeing. I chattered gaily as we walked up the lane together to the root ground.

"Won't take us long to hoe through that lot, will it, Fred? I wonder what we'll be doing tomorrow?" Fred looked at me and grinned.

"Us'll be on this old job fer more'n a week or two, I can tell yer. Flat-'oein' fust – that's goin' up atween the rows – then we got to single it all."

"What's 'singling'?" I asked.

"Well, you've seen mangol' seed, han't yer, Boy? Like little clusters o' seeds it is – all stuck together. They all grows – every last one o' 'em – comes up like a box 'edge, it do, up the row. If you was ter leave 'em be they'd strangle one-'nother an' grow all skimpy an' small. But you wants a gurt, big mangol' so you cuts out all bar the best-'uns, leavin' 'em 'bout fifteen inches apart, like."

"Sounds a lot of work, Fred," I said, the first doubts beginning to enter my mind.

"'Old 'ard, Boy," he said, "I aint done yet. When yer finished singlin' yer seconds 'em – that's flat-'oein' again ter cut out any fresh rubbish what's growed, an' cuttin' out the doubles yer left the fust time."

"But doesn't…?"

Fred continued remorselessly, "…then we goes back an' starts all over again, flat-oein' just ter make sure of un." He thought for a moment. "If we keeps at it, an' does a bit o' overtime on Sat'days an' Sundays, like – us *might* be done in three weeks."

"I didn't realise there was so much to it, Fred," I said feebly, "but it's an easy job isn't it?"

"There's a 'ell of a lot you don't know yet, Boy," said Fred. "but it'll be easy 'nough when you knows 'ow. An' don't you worry about that. I'll learn you proper, I will. I aint 'avin' no bad work done along of I." He paused to light his clay pipe.

"But us'll get on famous this year, wont' us? Wi' you bein' so keen an' all, us'll be up an' down them rows like greased bliddy lightnin', won't us?"

We reached the field and started hoeing. I set off at a great rate to demonstrate how easy I found the job. Fred let me go on for some time and then, when I straightened up to ease my aching back, made me go all over my ground again. "An' this time, Boy, cut out all the weeds, not just 'alf on 'em."

I soon adopted Fred's steady pace and found that I was doing better work and that it was easier on my back. But even then I was in agony. By noon I thought I'd been crippled for life, and when the hands of Fred's watch eventually crawled round to five o'clock, I was convinced I'd spend the rest of my days bent double as if I were looking for a dropped sixpence. And Fred made me work for another twenty minutes.

"Us won't do no real overtime, not on yer fust day at it – but I aint goin' ter leave no row 'alf done – you keep a-goin', Boy, 'til us reaches th' 'eadland."

I stumbled home beside the old man in an exhausted silence. As we neared the farm he said,

"You didn't do too bad today, Boy. You'll like it better when yer gets used ter stoopin'."

"I shall never get used to it," I growled.

"Back ache, do it, Boy?" he asked. I nodded.

"Keep a-movin', that's the answer. When you've ad yer tea do summat. Don't go restin' in no easy chair or that'll set an' give yer gyp termorrer, that will." He grinned at me. "Tell yer what, Boy – I'm goin' ter 'oe my 'llotment tonight, when I've 'ad me tea. Care ter come over an' give I a 'and fer 'alf an hour?"

Sometimes you can't help being rude.

"Not bloody likely, Fred," I said.

The only thing I liked about hoeing was that Fred would talk about himself. He had a fund of stories about the farms he had worked on, the men and horses he'd worked with, and what it was like in the old days. He told me that, years ago, a farm the size of ours – 120 acres – would have had two or three horse teams, each with a carter; two dairymen for the cows, and a couple of labourers. And in addition, there would be casual labour – women and boys from the village – for peak times like harvest, haymaking and hoeing.

"Us used ter do a deal o' 'oein' in them days. The corn an' all got 'oed. But it were all right when there was a big gang of yer at it – yer covers the ground an' you 'as a bit of a laugh an' a lark about, an' that do pass the time."

In my imagination I peopled the field with Fred's happy gang of smock-clad hoers. Their cheerful voices and the rasp of their hoes as they worked in the sun with us rang loud.

Then the vision faded and I looked up from my daydream. There were only the two of us, alone in a ten-acre field. Even then, thirty years ago, I had the feeling that the laughing *real* people of the land were being swept up and thrust into the drab uniformity of towns, leaving us few survivors to usher in the age of the machine.

And during the month of May, mechanisation came to our dairy herd. The boss had been talking for some time about expanding the herd, but there were two things standing in his way. The first was the high cost of labour with hand milking. It used to be reckoned that you needed one milker for every ten cows. With milk at 2/6d a gallon and a farmworker on £5 a week, each cow had to produce 200 gallons each year to cover her labour cost; and that's before you'd started paying for her fodder or the cost of any bought-in concentrates.

The picture is somewhat different today. At 1984 prices the same hand-milked cow would have to produce nearly 800 gallons to pay her milker's wages – and that's before she's been housed and fed.

The second, and most important objection to the boss's grand plans for what was eventually to become an efficient, one-man operated, 100-cow Friesian herd, was that our old, small, lean-to cowstall was filled to capacity. Putting up a new, modern cowstall or parlour was out of the question because of the high capital cost, so the boss decided to follow the example of A.G. Street, Rex Paterson, A.J. Hosier (the inventor of the outdoor bail), and many other farmers in the South of England, and start milking by machine, out of doors.

The Gascoygne bail was a four-stall, mobile milking parlour mounted on skids. Between each pair of standings was a churn with a vacuum-tight lid into which the milk flowed direct from the cow, and a hopper for concentrates. A chute from the hopper led down to a small manger at the side of each cow's head so that a ration of concentrates could be fed during milking – we used a home-grown mixture of rolled oats and peas called 'dredge corn', fed at 4lbs for each gallon of milk the cow was producing.

There was no floor to the bail so there was never any

cleaning out to do. When the ground got dirty you towed the bail to a fresh pitch with the tractor, and even in wet weather when you had to move after each milking, it only took a couple of minutes to do.

The four milking machine units were powered by a vacuum pump driven by a Petter 1½ hp petrol engine, housed in a cupboard at the drawbar end of the bail. The cows stood in the stalls, held in with a chain round their rumps, and were quickly milked out while they had their small feed. Then they were let out of doors at the front, and the next four took their places.

We started training the cows to the new system with the bail parked in the yard outside the cowstall. At first they had to be bullied, threatened, cajoled, and in some cases dragged by brute force before they would enter the strange machine, but they soon found out that there was a feed waiting for them, and that being milked by machines was more comfortable than hand milking. A machine sucks at the same vacuum level as a calf, the only difference being that it has four mouths all sucking at once, it doesn't have sharp little teeth like a calf, and it doesn't butt its head up into the cow's bag in the excitement of being fed. From the cow's point of view, something of an improvement on nature.

With all this luxury treatment, and the extra feed we gave them to keep them happy and entice them into the bail, the cows were soon clustered round it, eagerly waiting their turn. And we found that we had to alter our attitude to the cows slightly. There was no room for: 'I'm the cowman and you're the cow so when I says get in the bail you GOES!', because the cow will do the opposite and be off across the field like a long-dog. If you're milking on your own there's very little you can do about that. Also, cows being very sensitive to atmosphere, if you upset one you upset the whole herd, and they'll *all* stream off to the other side of the field leaving the cowman dancing with rage and talking about that place in Surrey called Effingham.

As the tone of voice is so important with animals, it is quite normal to hear a cowman relieving his feelings by berating a misbehaving cow. 'You're a rotten old bugger and I hate your guts,' he'll say. But he'll say the words with a smarmy smile on his face, and in a tone dripping with gratuitous loving-kindness. Perhaps another reason why observant townspeople think us country yokels are a bit funny in the head.

While we were training the cows to go in the bail we noticed an interesting example of their addiction to routine and dislike of change. The engine of the bail, a sturdy four-stroke, was delivered without its exhaust silencer. Someone in the Gascoygne factory at Reading had forgotten to put it on. The fitter who came with the outfit to set it up for us and get it going promised that it would be sent "As soon as poss., Guv!", but it took three days to arrive. During that three days the machine-gun clatter of the open exhaust added to the growl of the vacuum pump gave a noise level that made speech impossible. We managed to minimise it by piping the exhaust up under the eaves of the pig yard – rather to the astonishment of a sow and litter who were in there at the time – but they, like the cows, soon got used to the din and took no notice of it.

The postman brought the new silencer halfway through milking one morning, and we stopped the engine for a moment while the fitter screwed it on. He swung the engine and started it again, closed the cupboard door, and the exhaust noise was reduced to a subdued mutter. But the cows that had been waiting to be milked backed away from the bail and refused to go into it, and had to be coaxed in all over again.

"You're an ungrateful lot o' ol buggers," said Ernie as he held out a scoopful of oats and peas, "an' yer don't 'preciate nothin' we does fer yer!" But he said it in a wheedling tone, with his cowman's simpering grin fixed on his face.

Within a couple of weeks the whole outfit, cows and all, was out in the field and the cowstall stood empty and deserted. The boss walked about beaming at the speed of milking with machines, and talked gleefully of buying more cows to get the herd up to sixty head, which was the one-man capacity of a four-point bail in those days.

Fred as usual, thought differently.

"You mark my words, Boy," he said. "they cows'll be back in that ol' shed come winter. No-one in their right mind an't goin' ter be keepin' milkin'-cows out in th' field when it be snowin'."

That was the opinion of the expert critics in the village – that it was downright cruel to expect a man to milk outside in the cold – and that it was even crueller to make the cows live out all the year round. What they forgot (conveniently) to take into account was that nature designed cows to live outside in the winter. Man puts them into a warm shed because it's more economical to do so, if you can afford a big enough shed, and if you can afford to employ someone to clean it daily. The cows need a much lower 'maintenance' ration as they don't need to keep themselves warm – the shed does that for them.

But keeping cows outside, the extra cost of feeding is more that outweighed by the colossal saving in labour, and a bonus is the better health of the herd. Our vet, on one of his rare visits to the farm remarked ruefully to me, "Your damn cows are too healthy. If *everybody* went in for bail milking, I'd be out of blasted business!"

The same applies to the man who works on the bail. After leaving college I milked out of doors for six years and don't remember getting a cold in all that time.

The cows grew thick, hairy coats, as nature intended them to, and thrived in the cold. They'd lie there, cudding happily, with six inches of snow on their backs. One in particular was a good example of the healthy, outdoor life. She was a Jersey, a breed which was normally coddled and pampered by 'Gentleman Farmers' who delighted in a herd of the smooth-coated, shiny, fawn-like animals. Her name was Swappy, and she came from the farm belonging to

Yvonne Arnaud the actress, near Guildford, at the time when animal feedingstuffs were strictly rationed. It must have been some sort of 'Black Market' deal the boss did, because the little heifer was 'swopped' for two bags of oats – hence her name. She joined the herd of rough-coated Shorthorns and Friesians on the bail and never looked back. In winter she grew such a thick coat that she looked like a teddy bear and she milked very well. She had four heifer calves during her life – Penny, Tuppence, Joey and Sixpence – and they were all pure Jersey and as tough as their mother.

Self-appointed 'experts' – and a village is full of them – are the bane of a farmer's existence. Everything he does is in the full view of the public and is commented on with the benefit of hindsight; but without the responsibility of making quick decisions, the results of which will be obvious for all to see for many months, or even years.

If the farmer is haymaking, the hay isn't *quite* fit, there is rain forecast for the morrow, and he can't decide whether to go ahead and stack it, what is he to do? If he stacks it, it doesn't rain, and the stack heats because the hay was a bit green, the critics will say, 'The man's daft. I could've told 'im that hay needed 'nother couple o' days!'

And if he decides to hold back, the rain comes down the next day and the hay is washed out and ruined, it'll be, 'Bloody fool! Should've put un in th' rick while 'e 'ad the chance. 'Make 'ay while th' sun shines', my old dad allus said ter me. Farmers these days – they han't got no idea…' and so on.

When our outdoor herd proved to be a success the critics, again with the benefit of hindsight, congratulated themselves on their foresight. 'I allus said as 'ow that were a good idea. Them cows looks real well, don't 'em? Well – after all – 'tis natural ter keep 'em outside, an't it? Could've told 'im that years ago.'

The empty cowstall worried the boss as it was standing there costing money to keep up, and not earning anything. So he suggested to Fred that we could keep more pigs –

additional breeding sows and a stock boar. Fred was enthusiastic.

"If us makes pens in that old cowstall, wi' breeze blocks an' that – that'll be just right for my old pigs," he said. "Good job them bliddy old cows is up the field out the way."

The boss looked at me and grinned. Fred looked at me and winked. They each thought they'd won.

6

JUNE

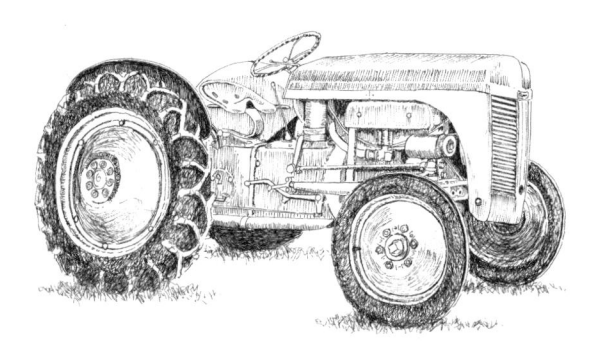

Fred and I finished singling and seconding the mangolds and marrow-stem kale at the beginning of June. I was pleased that we didn't have to single the thousand-head kale – we just flat-hoed between the rows. Fred, of course, wanted to single it, "Do the job proper, like us allus used ter" but the boss decreed otherwise. It wasn't worth the time, he said, the extra yield wouldn't cover the labour cost.

"It bliddy well could." said Fred to me when the boss had gone. "An' anyways – I an't on about costsies an' that – I'm talkin' about doin' th' job *right*."

"I agree with him," I said loftily, "except that it is now thought that singling thousand-head kale actually decreases the yield; an unsingled crop gives a greater total yield than one than one that has been thinned, and at least as high a proportion of leaf." (I was quoting madly from 'Watson and Moore', into which I dipped to try and find out about any new crop I encountered.)

"An' what the bliddy 'ell do you think you knows about it" said Fred angrily. He glared at me.

"My book…" I began weakly, but Fred snorted loudly, said "Bliddy book-learnin'!", turned his back on me and refused to speak for half an hour.

When he eventually recognised my survival from the terrible ordeal of silence, he said forcibly,

"I don't grow no kale in my 'llotment – but if I did – I'd bloody well single it!"

There was no argument, however, about the necessity of singling the couple of acres of fodder beet grown for Fred's sows. Fodder beet is a cross between mangolds and sugar beet, and I'd looked them up too. I was delighted to read that fodder beat has twice the feeding value of mangolds, and was going to have another go at Fred about the value of mangolds as a crop, when I read further that mangolds yield twice as heavily as fodder beet. So I kept quiet. I was sure Fred would shoot me down on that one, even though he 'never 'adn't 'ad no book-learnin''.

I wasn't looking forward to flat-hoeing all the roots again, but as it happened I was spared that torture.

"It's getting late," the boss said, "we shall soon be into silage-making. We ought to flat-hoe once more, but we'll have to make do with just horse-hoeing the ground."

I had a sneaking suspicion that there was more to that decision than met the eye. I don't think I was the only one to hate hand-hoeing. The boss came out with Fred and me a few times but, strangely, only on the days the corn merchant's traveller was due. The two of them stood talking prices for hours, while Fred and I scraped our way up and down the rows, then, just before dinner time, the boss came back and said, "That bloke! Talks the hind leg off a donkey, he does. Thought I'd *never* get away!"

I told Fred I thought it was all prearranged to get the boss out of hoeing.

"Yer on'y just noticed that, Boy?" He touched the side of his nose knowingly. "But 'is leisure time do cost 'im summat. Seen 'ow much fert'lizer an' cow cake an' that we been getting' lately?"

And Ernie said how disappointed he was that hoeing was finished.

"I were lookin' forrard to a bit o' that, Fred," he said in an aggrieved tone, "but you an' Boy done it so quick this year I never 'ad no chance."

"Lyin booger," said Fred. "I see you a-peepin' o'er the 'edge at us when me an' Boy was bendin' our backs. Yer never did no volunteerin' then, did yer?"

To be fair, a good part of the boss's absence from the root field could be explained by the need to keep an eye on the handyman from Gomshall, who was busy putting breeze-block partitions in the cowstall. One of the first jobs he did for the boss ended in disaster. He was commissioned to build a new double pig sty – the traditional type, with sloping-roofed shelters, each with an outside run – in the paddock behind the farm. He was a good builder, and the breeze-blocks were laid straight and true, but he knew little about pigs, particularly their strength.

On the day the new sties were finished, we had a sort of launching ceremony. The inside shelters were bedded-up ("Always use short straw for sows that are going to farrow," the boss told me. "Little piglets get tangled in long straw and can't get out of the way when the sow lies down – they get crushed."); new, galvanized hinges and bolts were fitted to the neat little doors the handyman had made; and, to add the finishing touch, we put a pig in one of the sties.

She was a pedigree Large White the boss had bought as an in-pig gilt a few days previously, and she was the apple of his eye. She had a long-winded pedigree name which was much too much of a mouthful for everyday farm use, so the boss, who gave nicknames to everything and everybody on the farm, said that henceforth she was to be known as 'Gorgeous Gussie'. This was after the long-legged, frilly-pantied tennis star Gussie Moran, whose picture was in all the papers at the time.

Gussie the pig walked eagerly into the sty, went into the shelter and scuffled the straw about, then stood in the doorway looking at us expectantly as if to say, 'What happens next? Food?', and when no food materialised she decided to celebrate in her own way by having a good scratch against the wall dividing the two outside runs.

When a pig scratches she does it with energy and dedication – literally throwing herself into the job. Gussie had just got a good stroke going – leaning against the wall at forty-five degrees, the rough, bristly skin of her back rasping against the coarse breeze-blocks, eyes tight-closed and a look of blissful pleasure on her screwed-up face – when the calamity occurred.

There was a loud cracking sound followed by a crash as the whole dividing wall fell over in one piece before disintegrating into its constituent blocks on the floor of the next pen. Gussie went over on top of the blocks, uttered a loud squeal of terror, scrambled to her feet and dashed into the shelter of her sty. She stayed there, not daring to emerge for about twenty minutes – not so much because she was frightened of more walls falling about her ears, but she could hear the boss telling me in great detail the trials and tribulations that beset poor farmers.

What was the good, he said, of trying to make an honest living when he was surrounded by incompetent idiots and animals bent on destroying everything in sight? And further, why was it that *everyone* had to be supervised *all* the time when they were doing *any* bloody thing? I ventured to suggest that six-inch blocks would have been better than the four-inch ones that had given way, and the boss agreed. Why was it, he asked me almost tearfully, that he had to think of all these things himself? Couldn't anybody else do *anything* right?

He definitely wasn't pleased, but at least he had a good reason for staying away from the root hoeing, and the block walls in the cowstall were made of six-inch blocks and were good and strong – able to withstand the ravages of time, destructive pigs and heavy-handed farmworkers.

More sows went into the new pens – some bought-in

Large Whites, and some Wessex Saddleback gilts that the boss had kept for breeding from a litter of black and white weaners he reared – and one pen was reserved for a young stock boar.

Previously the boss had taken our sows to a boar belonging to Tony Reid, a pig farmer in Shere, but when the number of sows increased this became too wasteful of time. So Sammy, the boar, arrived and was duly installed.

At the start, Sammy was a bitter disappointment to us. He looked normal, he ate the food that Fred gave him normally, he snored in the normal manner of a pig on the bed of straw he rucked-up for himself in the corner of his pen; but he flatly refused to serve the sows that were put in with him when they were 'hogging'.

One morning Fred asked me to help him with a sow.

"One o' me sows be 'oggin', Boy. She'm well on. You give I a 'and an' we'll put 'er in wi' that silly young booger – see if 'e'll 'ave a goo this time."

We put the sow in Sammy's pen and she made the usual advances to him, rubbing against him and grunting expectantly. Sammy took no notice of her at all but just stood there, a faraway look in his eyes.

Ernie came striding along the passage, on his way to feed some calves. He joined us in the doorway looking thoughtfully at the two pigs.

"Know what, Fred?" he said. "I reckon you give that there boar the wrong name."

"Nothin' wrong wi' the name, Ern, Sammy's a good name fer a pig. Allus calls a boar-pig Sammy round 'ere, us does. Allus worked afore, they 'as."

"Not fer that one it ain't." Ernie dug Fred in the ribs. "Reckon 'e ought to be called Nancy, I do."

The next one to join the group was Jill Newman, the boss's wife.

"No luck, Fred?" she said. "What's wrong with him?"

Fred touched his cap.

"Dunno missus. Can't make it out, I can't."

We stood there in thoughtful silence, each one wondering how we could instruct Sammy in the mysteries of what should come naturally to a young male. I had no experience in the job, but thought that if a comely young lady flattered me with the attentions that were being given to Sammy by a female of his species, I would respond a lot more eagerly than he was doing, which was not at all.

The boss's wife spoke.

"I've been thinking," she said. "maybe we shouldn't all be standing here staring at them. Perhaps he doesn't like being watched."

"Ar!" Fred nodded his agreement. "Damme if I does neither."

There was no answer to that, and the meeting broke up leaving the sow and boar together. We left them there for an hour or two but had to hurry back when squeals of protest from Sammy indicated that the disappointed sow had turned on him with her teeth.

The boss was in a quandary about the boar, and got the vet to look at him. He was pronounced fit, and all his equipment appeared to be in working order, so we were just as puzzled as before. Then someone suggested the old pigman who worked for Mr Mathews at Manor Farm, Wotton – if he didn't know what to do, then nobody would. The boss went off to fetch him in his car.

The old man climbed stiffy out of the Ford.

"Can''t stop long," he said. "I be in a 'urry today."

We showed him Sammy who was as usual standing in his pen staring moodily into space.

"Clear up 'is grub, do 'e?" he asked.

"Oh yes." replied the boss. "No trouble that end – it's the other end I'm bothered about."

"Well you'm wrong, Guv'nor. 'Tis in 'is 'ead, what's wrong wid 'e. 'E's fed up an' mis'able. Yer can see by the look in 'is eye. 'Ave yer tried un in a different pen?"

"We thought of that and tried him in every pen in the yard. He's the same in all of them."

The pigman nodded wisely.

"Well – there's only one thing for it an't there? – an' yer knows what that is, don't yer?"

"Yes, I do," said the boss wearily, "shoot the bloody thing and get another young boar."

"'Old 'ard – I never said that, Guv'nor. Like I said 'e's fed up – wants summat to occupy 'is mind. You get a nice, shiny tin can – make sure it an't got no sharp edges on it – an' chuck it in 'is pen for un ter play with. That'll do the trick, that will." He stumped off towards the car.

"Come on, Guv'nor, you get I 'ome. I got a heap o' work to do today."

We did as we were told and threw a tin can into Sammy's pen. He eyed it suspiciously at first, and then started playing with it, pushing around the pen with his snout, picking it up in his tusky mouth and shaking it like a dog with a stick, and throwing it round the pen. For the next few days – except at feeding time when there was a busy silence – we could hear the clanking of Sammy's new toy.

The first time Fred put a sow that was hogging in with the now contented young boar, he came running to the boss with the joyful news.

"Thank God for that!" The boss was delighted. "He's working, is he, Fred?"

"Workin'?" said Fred. "I never 'ad no time to shut the door afore 'e was up on 'er an' goin' like a steam-engine. Where be that Ernie? I'm a-goin' to tell 'im 'bout my Sammy. I'll give 'im bliddy 'Nancy', I will!"

Horse-hoeing was different altogether to hand-hoeing. I led Kitty, the mare, up and down the rows at a steady walk while Fred steered the single-row hoe. It was a satisfying job as, after the slow, back-breaking job of hand-hoeing, the amount of ground we covered in a day seemed enormous.

There was only one snag to the job – horseflies. They were attracted by the glossy, sweaty, coat of the horse and we worked in a cloud of them. It wasn't too bad for me – leading the horse I had one hand free to slap at the horrible things – but poor old Fred couldn't defend himself as he had to keep both hands on the handles of the horse-hoe.

"Damn an' blast they old Stout-flies," he swore ('Stouts' was the local name for horse-flies), "they do fair tug at yer. The 'oss be all right – she been fitted wi' a tail fer swattin' 'em – but I an't got nothin'."

Stouts weren't the only pests to bother us in the root ground – there was another insect which, although it didn't suck blood, was far more dangerous to the farmer.

I had just turned the mare at the headland and was walking dreamily up yet another row when Fred's shout, 'Ho-Hup, Boy!' pulled me up short. I turned, expecting to see that Fred wanted to fill his pipe, or the horse-hoe had come apart – and saw him leave the handles of the hoe and walk quickly over to some newly-sown kale. He stooped over it, examining it closely. I went over.

"What's the matter, Fred? What is it?" He pointed at the tiny kale plants, just coming through.

"That's what be the matter, Boy. This 'ere kale 'ave got the fly. Look-ee there! See 'em jump?" There were hundreds of tiny black insects leaping about in the rows.

"They're like fleas, Fred."

"That's why they be called flea-beetle," he replied.

"But I thought you said they were called fly," I said, wanting to get it straight. Fred glared at me.

"Don't stand there arguin' about what they be called, Boy, you run down the farm an' tell boss as we got 'em. We got ter do summat about they real quick, we 'ave."

I trotted off down the lane to the farm, impressed with the urgency of the situation. The boss was talking to Ernie in the yard.

"Fred says we've got fleas or fly-beetle or something on the kale. He says we've got to do something about it." I didn't think I'd got it right but the boss responded immediately.

"Ernie, find me a ten-foot pole, and I shall want some of your cake bags, and some string to tie them on with." He turned to me. "Don't stand there gaping, go and fill two five-gallon cans with T.V.O. I'll get the tractor out."

By the time I'd filled the two cans with paraffin the boss had the Fordson warmed-up and hitched to the trailer. Ernie piled sacking, string, a tow-chain and a long pole on it and we climbed aboard. Directly we were on the boss let in the clutch and we roared off at the tractor's maximum speed – eight miles per hour.

"What's all the rush for, Ernie?" I asked as we careered along, bouncing uncomfortably, hanging grimly on to the cans of paraffin, trying to stop them spilling.

"'Cause we wants a bit o' kale next winter." he replied. "If we doesn't stop they little devils wi' a dose o' paraffin they'll nip that kale off clean as a whistle, they will."

We got to the field and the boss quickly unhitched the trailer. He chained the pole behind the tractor like a harrow pole and we tied the sacks all along it. The sacks were doused liberally with the paraffin and the boss drove up and down the rows of kale so that each plant got a good soaking. We had to keep on re-wetting the sacks but eventually the job was done.

I asked Ernie if paraffin would harm the kale – the plants seemed so small and fragile – and whether it was a sure cure for the fly.

"No, it won't do it no 'arm," he said, "an' you look an' see if you can find any fly now we been over it."

I looked – and reported that the beetles had disappeared like magic. I wondered where they had gone.

"Off down to old Fred's 'llotment I 'spects." he said. "You stop an' 'ave a listen when you passes by tonight. 'Ear 'em chompin' 'is cabbage, you will. Ar an' spittin out the tough bits."

"Yer won't find no fly on my greens, Boy," said Fred, "an' yer won't find no weeds down th' rows neither. 'Tisn't like a cowman's garden – all weeds an' no crops."

As we went back to the horse-hoeing I asked Fred how he made sure his young cabbages weren't attacked by the fly.

"Cause I takes a bit o' trouble wi' it, Boy." he replied. "When I sows cabbage an' that I pinches a bit o' paraffin out the tractor shed an' doses the seed wi' t before I sows un. That stops that old fly from comin' near it."

I asked him why we didn't do the same with the farm crops and he explained patiently that seed wetted with paraffin wouldn't run through the root drill. So you had to wait until the beetles attacked before dealing with them. In subsequent years we used a systemic dressing – a powder mixed with the seed just before drilling. It was completely effective, but I always had a sneaking feeling that it was unfair on the poor little beetles. They had no chance.

While we'd been hoeing and horse-hoeing, the boss had hired a dragline excavator from the County Agricultural Committee – the 'War-Ag' as we still called it, five years after the war had ended. Silage pits were dug at strategic points around the farm and we were proud that we were the first farm in the area to make silage. Fred, of course, was dubious.

"I don't reckon much o' silage." he said. "Us used to make it, years ago, if us 'ad a bit o' 'ay what was bein' spoilt by th' weather – but it never turned out no good. No guts to it."

The boss explained that making silage out of half-made hay that had already been spoilt was a waste of time. The silage we were going to make was put into the pits green and fermented properly. He'd read all the latest books on it, and the leaflets put out by the Ministry experts, he said, and knew it was the right system to use. Fred said we'd see, wouldn't us. If he didn't go much on silage he went even less on "they Min'stry fellers an' their bliddy books."

The first crop we ensiled was a massive cut of oats, peas and tares – arable silage. Urged on by a good coating of dung and a top-dressing of 'bag muck' the peas and tares had grown luxuriantly, twining round the thick-bladed oats and themselves to form an impenetrable mass. The old McCormick horse mower, towed by the Fordson, cut it off all right, but left it standing as before with no trace of where the cutterbar had passed through. We had to tear each swath away from the standing crop with prongs before we could see where to make the next pass with the tractor and mower.

Fred remarked mildly, "'Tis a 'agglin' old job an' no mistake."

Once we'd cut enough to cart in one day we hitched the four-wheeled trolley to the Fordson and loaded the greenstuff on to it by hand. I have never, since, encountered anything so heavy and unmanageable as that oats and peas. It clung together so tightly that we swore you could have hitched a horse to one end of the swath, shout 'Git-Hup!', and the whole thing would have towed to the silo unbroken.

We cursed and sweated and broke prong handles on it. I began to wish that I'd run away to sea instead of choosing farming, and was having an easy time of it battling with gales round the Horn; or that silage hadn't been invented.

We had all sorts of troubles. We put too much on the first load and the small, iron wheels of the trolley sank into the ground until the Fordson couldn't budge it – we had to unload half to get it going again; and the second load, when we got it to the pit, sank into the soft mass of the first and became equally immoveable.

At the end of the day we had managed to clear all the greenstuff we'd cut, but at a great cost in broken prongs, aching muscles, many cuss-words and a great deal of time. But we had learnt a lot – after that we used a small, rubber-tyred, two-wheeled trailer and took smaller loads more often.

And the best result of all was that the boss, taking into account the enormous amount of labour involved in making silage by hand, decided to buy a Ferguson tractor.

For some time I had been casting an envious eye over the hedge to where our neighbour's tractor driver was using a Ferguson. The sight of the little grey tractor purring up and down with its mounted implements, while we struggled with the cumbersome Fordson, made me think that we were getting behind the times; but the boss had a few arguments against buying a 'Fergie', but the main one being the money involved.

"Oh, it looks good," he said, "but that one runs on petrol which costs three times as much as T.V.O., and they're expensive to equip. We got all trailed machinery, which wouldn't do for a Ferguson, so we'd have to buy a new plough, a cultivator, a special trailer and God knows what else. It'd cost about a thousand pounds in all – you could buy twenty good cows for that."

I understood what he meant. One thousand pounds was an enormous sum – about a quarter of the price the boss had paid for the whole 120 acre farm. But some crafty salesman must have told him about a machine called a buckrake, and our experience with the arable silage did the rest. Within a few days a brand-new, paraffin-engined Ferguson was on the farm complete with its two-furrow, mounted plough and a Paterson buckrake.

The buckrake, invented by Rex Paterson, a silage and outdoor-bail enthusiast, was, I think, the greatest step-forward in low-cost fodder conservation ever made. Like all brilliant ideas it was very simple – like a small-sized haysweep mounted on the hydraulic lift of a modern tractor. You reversed along the swaths of cut greenstuff with the tines on the ground gathering up the crop, and when you'd got about half a ton on the buckrake, you lifted it up and carried the load to the silo. When you got there you ran up onto the heap, dumped the greenstuff by pulling a trip-lever and went back for another load. Very fast and efficient. The only person who complained was Fred, spreading out the grass on the silo with a prong.

"Fer Gawd's sakes, Boss, goo stiddy. I be fair snowed under."

We used a natural additive on the silage – thick, black molasses put on with a watering can at one gallon, mixed with two gallons of water to every ton of greenstuff. That was usually my job and I ended up most days with a fair covering of the filthy black stuff all over my clothes and boots. To begin with I dipped my fingers in the barrel and licked them but Fred warned me about that.

"Don't you take too much o' that old treacle, Boy. Give yer th' the trots, that will. We wants yer ter do some work, 'stead o' runnin' be'ind th' 'edge all day long."

Once we'd got the hang of it, silage-making went on apace. Silage is greenstuff pickled in naturally-produced lactic acid and the heap has to be maintained at the right temperature – about blood heat – so that the correct fermentation can occur. The boss talked knowingly about the optimum temperature for lactic-acid bacteria and Fred was mystified.

"What the 'ell's 'e on about, Boy? All that stuff 'bout lax'tives an' that? An' us never used no bactrians in th' old days – just molasses. I 'spect 'e bin readin' them books again. Don't yer take no notice, Boy, just keep th' 'eap 'and-'ot an' us won't goo far wrong."

It was also my job to keep the temperature of the silo correct and I did this by rolling the heap with the Fordson. It was rolled in the mornings before we started adding to it, several times during the day as each layer went on, and finally last thing at night. If the temperature started to rise, it was cooled by rolling, and if it dropped too low we added greenstuff and rolled less.

One morning Fred and I arrived at the silage pit, and while Fred busied himself with the churns of water we'd brought to dilute the molasses, I drove the Fordson up onto the silage. I stopped short because there, curled up in the warmth, was a snake. I got off the tractor, looked at it and then called Fred.

"What be up *now*, Boy?" he grumbled as he climbed up the heap. Then he saw the snake and jumped back. "'Tis a bliddy deaf adder – keep clear, Boy!"

I didn't need telling twice. I hate snakes and the fact that this one was venomous made it worse – its cold, reptilian threat sent shivers down my back. I kept about thirty feet clear.

Fred fetched his prong and killed the thing expertly with one blow behind the head. Then he picked it up gingerly over the tines of the prong and hung it on the hedge. Its reflexes made it writhe slowly.

"It isn't dead!" I said, feeling sorry for the creature despite my loathing.

"Yes 'tis," said Fred, "leastways, 'twon't do no 'arm now. They says as 'ow a snake don't die 'til sunset yer know, Boy."

I watched the snake all that day, touching it from time to time with a long stick. Each time I did so its body responded with a small movement. But at sunset it was still.

7

JULY

The brand-new Ferguson and its directly-mounted buckrake had been worth their weight in gold when we were silage-making – we were all agreed on that – but when I enthused about the wonders of the little grey tractor, Ernie and Fred turned severely on me and warned me not to speak too soon. 'The proof o' th' pudden be in th' eatin', Boy' they said, and they meant it very seriously. Any new machine, man, animal or farming practice had to be tried and tested (preferably on a neighbour's farm) before it was fully accepted. They weren't saying that the Ferguson wasn't any good – nobody would have been fool enough to criticise that marvellous hydraulic lift after it saved us so much back-breaking work – it was just that the tractor hadn't proved itself on ordinary farm work. And its slim twenty hundredweight against the Fordson's ton and a half, and the quiet purr of Fergusons's nineteen hp overhead-valve engine compared to the noisy bark of the twenty-five hp, old-fashioned side-valve, made the new tractor seem like a toy.

"They don't make 'em like that any more." Ernie patted the old tractor's cast-iron radiator. "Them new things is made o' tin."

Ernie's implied criticism of the Ferguson's lightweight and modern styling was correct in one respect – they certainly didn't make tractors like the Fordson any more. Fordsons were simply machines which dragged trailed implements through the ground by brute force – implements which hadn't changed in basic design since the days of ox teams. Indeed, the only implement in use at Hatch Farm specifically designed for the Fordson was a two-furrowed trailer plough, and that could have been towed by a horse team without too much trouble.

What Ernie didn't realise, and neither did anyone at the time, was that we were witnessing the biggest change in farm machinery design since the ard – the earliest form of plough – took over from the digging stick.

Harry Ferguson's invention, the 'Ferguson System' was not merely to put the tractor and plough together as one integrated machine instead of two separate units, but to do away with wheels on the plough, controlling the working depth hydraulically. Thus the plough was very much lighter but the most important thing of all was that all the weight of the plough, plus the 'suck' of the shares into the ground, was transferred directly to the back wheels of the tractor, giving it traction without a lot of built-in weight or added ballast.

A good example of the relative efficiency of the two tractors was the fuel consumption. The Fordson, pulling two furrows, would plough three acres a day and use seventeen gallons of T.V.O. A Ferguson, pulling a three-furrow mounted plough, would plough five acres in the same eight hours and use only eight gallons of T.V.O.

But trying to explain or forecast a fifty per cent increase in output, coupled with a fuel saving – in T.V.O. at 10d a gallon – amounting to slightly than half the tractor driver's wages for the eight-hour day, was impossible without a practical demonstration. For generations, farmers and farmworkers had been concerned with breeding heavier and more powerful working horses. Their first introduction to mechanical power on the farm – still within living memory to some of them – had been the replacement of the simple flail by the threshing machine and a steam engine weighing more than twenty heavy horses put together. Harry Ferguson's tractor looked the size of a Shetland pony and weighed about the same as our old cart mare, Kitty. To suggest to Ernie or Fred that it would out-perform a team of six horses was a well-nigh impossible task. They needed a convincing show of proof on our own ground.

We had a day or two to spare before we started haymaking and the Ferguson was put to the test, ploughing the hard, dry ground the arable silage had come off. Well – we tried anyway.

The day didn't start off too well. We went to get the tractor out of the shed and discovered that one of the back tyres was flat. We tried to get the wheel off but the wheel nuts seemed to have been tightened up by some superman at the factory in Coventry – we couldn't budge them at all. The boss didn't waste time.

"Pump the tyre up hard and it'll stay up long enough for you to drive to the garage," he said to me, "they've got all the tackle up there to get the wheel off and mend the puncture."

After ten minutes with a foot pump I drove as fast as I could, up through the village to the garage opposite the post office. As I rounded the corner by the Abinger clock I felt a twinge of uneasiness when I saw the figure of Mr Bishop, the village policeman, standing outside the post office. The uneasiness turned to sheer terror as he stepped off the pavement into the road, his raised right hand commanding me to halt.

I stopped in front of him and waited for the sky to fall on me. I had no licence, no 'L' plates, and the tractor had no number plates or road fund licence. The words 'threw the book at him' came into my mind and I sat there trembling.

The policeman grim-faced and serious, walked slowly up to the shiny new tractor.

"That looks a very nice little machine you got there, Lad. New, is it?"

I gulped and nodded my head, unable to speak.

"I aint seen one o' these close-to, before."

He walked all round the Ferguson, slowly, so as not to miss any of the details. Then he walked slowly back again the other way. When he got back to his position in front of the radiator he looked at me sternly and said,

"And where might you be off to in such a tearin' 'urry?"

I found my voice with difficulty.

"I'm going to the garage, Mr Bishop," I croaked, "we've got a puncture and it's going down fast."

"Ar, you best be off then, han't you, Lad? But before you goes..." he stooped and flicked a speck of mud off the knife-edge crease in his trousers, "...don't you forget to see about them number plates an' that. You can't go about on the King's 'Ighway like that, you know – 'taint legal."

When the puncture had been repaired I drove back to the yard and was exultant about my escape.

"I fooled him," I said.

"Don't you be so sure about that, Boy." said Fred. "Old Bishop's cleverer'n that. I were 'avin' a pint wi' 'e, t'other night, an' 'e were on about 'ow you was a-drivin' on the road wi' no ticket. You never fooled 'im – 'e were a-warnin' yer."

I went straight to the post office and applied for a provisional licence. It cost me five shillings – half my week's pocket money.

The boss and I took the Ferguson with its new two-furrow plough up to the field to try it out – and it was a disaster. The plough wouldn't penetrate the hard ground properly, and when we did get it to go in, the tractor spun and skidded and reared up, but wouldn't go forward at all. We consulted the instruction books and tried every adjustment listed, plus one or two of our own that Harry Ferguson had never thought of, to no avail.

The boss was furious. He'd spent over three hundred pounds, he said, on a tractor that was useless. "Take the damn thing back to the yard and get the Fordson out. I'm going to ring Ben Turner's in Dorking and tell 'em exactly what they can do with Ferguson tractors," he said as he went off to ring up the local dealer.

I felt very sad about the failure of my pet but didn't get any sympathy from Ernie and Fred.

"Told yer they wasn't man enough, Boy," said Ernie. "All tractors is the same," said Fred. "You can't beat a bliddy good pair o' 'osses."

The boss came down from the farmhouse and he still looked gloomy.

"I phoned that chap at Turner's about the trouble we had and he wouldn't listen. Kept on about the 'Ferguson System'. And do you know what they're going to do now?" He looked helplessely at us, "They're going to bring a third furrow and stick on the plough!"

"An' you got ter pay for un, Boss?" asked Fred. The boss nodded glumly. "Then that'll be the system they uses – makin' yer put yer 'and in yer pocket."

The boss went back to the house after telling me to let me know the moment the man from Ben Turner's arrived. He wanted a word with him, he said.

The carefree confidence displayed by the salesman who bought the parts to convert the plough to three furrows did nothing to dispel the gloom in the farmyard. He whistled cheerfully as he tightened-up bolts and we watched him. We speculated on the size of the pieces of salesman that would be left after the boss had finished chewing him up.

"Cocky little devil," said Ernie, "an 'e must be daft as a brush ter think that a tractor what won't pull two furrows is goin' ter go wi' three."

The salesman heard him and grinned but said nothing. He went on whistling and winked at me. I had a sneaking suspicion that he was enjoying himself immensely – it was obvious that he'd been in the same situation before and had always emerged triumphant.

When the modifications were complete – the plough half again as long now, and looking enormous to anyone accustomed to two furrows behind a tractor or a single furrow behind a pair of horses – the salesman asked whether he could try out the tractor and plough on the same ground we had tried before.

"Of *course* we're going on the same ground." said the boss, "that's the whole point isn't it? And if that damn thing still isn't any good, you can take it back with you and I'll go and buy a Fordson Major."

The Feguson salesman frowned at such a terrible blasphemy and almost crossed himself, but nodded his agreement. We all trooped off to the field to watch the fun.

"Where do you want me to start, Sir?"

The boss pointed to the scratches and skid-marks we had made in our first attempts.

"Right there," he said. "Now get on with it."

The man drove to the place indicated and then stopped, with the tractor ticking over quietly, while he adjusted the top link of the hydraulic lift.

"You must get the top link set exactly right, it's the heart of the whole system." The boss's face was stony.

"We tried every setting in the book this morning."

"Maybe you weren't lucky enough to find the right one, Sir," said the salesman, then got hastily on to the tractor seat when he saw the look on the boss's face. He put the tractor in gear, opened the throttle and let in the clutch. The Feguson purred away across the field cutting three neat furrows.

"Well I'm damned!" said the boss.

"Gor bliddy 'ell!" said Fred. "Look at un goo!"

"I knew it would work," I said smugly. They both glared at me.

"Orrible bliddy liar." said Fred. "You was just as fed up wi' un as we was, 'smornin'."

But it was Fred the horseman who put the final accolade on Harry Ferguson and his tractors, a few weeks later. I overheard him talking to one of his cronies over a game of dominoes in the pub.

"You wants ter tell that boss o' yourn ter get one o' they little Fergies," he said. "They aint so big as t'others, but by God they does some work!"

The grass from the leys had been made into silage, and during July we started haymaking on the later-maturing meadows and permanent pastures. Although we were going to feed the cows mainly on silage the following winter, we still needed some hay. It is essential for calves and useful for sick animals, and Fred would have been most upset if anyone had suggested giving silage to Kitty, the mare. Another advantage that hay has over silage is that if you have any left at the end of winter you can keep it until the following year or even sell it as a cash crop. Although, if you did sell hay, you had to make it well known in the locality that it was a genuine surplus, or that the deal had been planned well in advance, as it was regarded as selling fertility off the farm – the act of a farmer in desperate financial straits. Hay fed to one's own stock goes back into the ground in the form of farmyard manure and keeps the land 'in good heart'. In the time before the farming slump of the Thirties, when the emphasis was on good husbandry and improvement of the land rather than reaping the highest possible cash output, the sale of hay or dung off a holding by a tenant was expressly forbidden under the terms of his lease. The tenant had to get his landlord's written permission to do such a thing, and if he did it without permission he would be evicted and then heavily penalised in the way of 'outgoing dilapidations'.

Of course there have been many farms and estates where the sale of hay was a legitimate and recognised part of the rotation, and a very profitable part. Before the internal combustion engine took over, all deliveries of goods in the towns and cities were done by horse-drawn vehicles, and vast quantities of hay were sold off farms to provide fuel for the horses. In 1919 the 'London price' of hay was £13 a ton when a top-grade farmworker was paid £2 a week. In the Fifties, the price of hay was about £7 a ton and the farmworker got £5 a week.

Some of the meadows – the steep hillside of 'The Hangers' in particular – had a grass with the local name of 'Pin-Wire Grass' in them. It was well named. It was almost impossible to cut when dry as it jammed the reciprocating knives of the old mowers. It had to be tackled when there was a heavy dew on the ground and the grass was soaking wet. We used to start mowing at four o'clock in the morning and keep going until the sun dried the dew off at about ten o'clock. I rode on the old McCormick mower, originally designed for a pair of horses, while the boss drove the tractor.

One misty early morning we started mowing and hadn't gone more than fifty yards when the mower ran into a big lump of wood hidden in the grass. I didn't have time to lift the cutterbar with the big hand lever, and the connecting rod or 'Pitman' that drove the knife broke under the strain. On the McCormick mowers the pitman was made of wood so that it formed a weak link. In the event of a blockage of serious proportions the wooden rod gave way, which was cheaper than the drive gears being damaged.

The fact that the wooden rod had performed its designed functions in life and given way under stress didn't comfort the boss. He delivered a long and lurid speech describing the parentage of the mower, the grass, the offending pitman rod, and last but by no means least the lump of wood. He hinted darkly as to what terrible fate awaited the idiot who had maliciously left it in the path of the mower and I kept very quiet – it might have been me. And while the boss did one of his ingenious repairs with baling wire I thought how lucky it was that I hadn't been driving the tractor – I would have been entirely to blame for not having seen the obstruction before the mower ran into it.

The art of making good hay is to dry the grass as quickly as possible after cutting it, so that the sugars and starches which are a large part of the feeding value are preserved. So the hay must be turned and 'tedded' – fluffed-up into a light, airy swath – to let the wind dry out the sappy stems. The mower leaves the grass in a tight, flat layer close to the ground and if it isn't moved the sun will bake up the top layer, leaving the underneath as wet and green as ever. 'Make hay while the sun shines' doesn't mean that you let the sun make the hay for you – it means that you must get the hay made and stacked before too much hot sun 'do bake the guts out've un.'

Later in the day Fred and I went to the top of the farm to get the horse-drawn swath-turner out of its winter quarters – a bed of nettles in the corner of last year's hayfield.

"Wouldn't it have been better to have stored it under cover?" I asked.

"Ar, it would an' all, Boy," Fred replied. "but 'er's a gurt, ock'ard thing ter put away – us han't got no room fer un in cartshed, or summat else'd 'ave ter stand out."

We pushed and heaved at the big wheels of the swath-turner, dragging it out of the clinging grasp of the nettles and turning it round so that we could put Kitty in the shafts.

"Don't you worry about un, Boy." said Fred. "Give un a

good oilin' an' 'er'll be all right – 'as been this last twenty year or more."

And so it proved. A quick run over with the oil can and the swath-turner worked as smooth as silk, the four rake bars with their wire tines turning with the precision of connecting rods on a steam locomotive, flipping the mown swaths over and exposing the green undersides to the wind and sun.

"We'll turn it again at tea-time," said the boss, "then put the middle tines in the rake bars and side-rake it – two rows into one – just before the dew comes down."

"Will it be ready to cart tomorrow?" I asked, knowing that side-raking or 'rowing-up' was usually the last operation before carting the hay.

"Off course it won't." The boss picked up a handful of grass and twisted it up into a tight rope. "See the knots in the stems, all bright green and sappy? When you can twist it up like that and it stays dry in your hand it's fit to carry – this'll be another three or four days yet." Fred nodded his agreement.

"An' that got ter be fluffed about an' tedded a good bit afore it'll be safe ter put un in th' rick."

"And the swath-turner does all that?" My affection for the ancient Lister-Blackstone 'Ted-Rake' grew as I contemplated a leisurely jaunt round the field two or three times each day – me sitting on the cast iron seat, controlling the mare with a casual flick of the reins. Fred looked at the boss and they both grinned.

"You tell him, Fred," said the boss.

"Yer won't goo far in farmin' if you thinks 'tis all a-sittin' on yer arse on 'chines all day long, Boy." said Fred. "We got special tools for teddin' – not like they tractor things what knocks all the leaf off – they'm called prongs, an' yer takes one in yer 'ands an' walks along the row fluffin' it proper like us used ter in th' old days."

"But I tell you what, Chris," said the boss generously, "as you're so keen on riding, we'll let you row the hay up each night with the side-rake and old Kitty. Fred and I'll find something else to occupy us while you're doing it."

"Like what?" I asked suspiciously.

"Like a pint of shandy with ice clinking round the glass."

"Best idea I 'eard in a long while," said Fred.

"On'y I likes me pint o' beer. Shandy han't got no bite to it when you'm a bit dry,"

For the next few days the three of us – and Ernie as well, when he could find time in between milkings – went out into the hayfield with prongs and spread the neat rows the side-rake had left the night before. We usually started about nine o'clock when the sun had dried the stubble between the rows.

"It's no use starting before the dew's off." said the boss. "If you spread hay onto damp ground it stays moist for hours. That's why I like to row it up at night. Some folk leave it spread overnight but then the hay keeps the sun and wind off the ground and it doesn't dry up until mid-day. Hay won't dry unless the ground underneath it is bone dry, and my way of doing it you gain three or four hours drying time each day – and that's important. The longer you leave hay on the ground, the more chance there is of it being spoilt by rain."

"What do you do if you think it's going to rain and the hay isn't dry enough to stack?" I asked.

"First you kick yourself for having mis-judged the weather," he replied, "and then you get busy and put it into cocks – like little stacks all over the field. If they're properly built they shed the rain and keep most of the hay from getting too wet. Then when the ground dries up again you spread it out to dry."

"And if the weather's catchy?"

We were working along a row each, keeping together in line, and Fred stopped to re-light his clay pipe. The boss and I stopped as well and waited for him. It's best to work as a team on jobs like that, each one encourages the others and companionable chatter makes the job easier.

Fred puffed out clouds of rank smoke and then carefully broke the match between his fingers to make sure it was out before throwing it down.

"Boogered if that boy don't ask a lot o' questions, Boss." he said, then looked at me. "If the weather be catchy, yer cocks yer 'ay up an' spreads un again – then yer does it all over again 'til it be fer to stack. Four or five times I've knowed it done – but yer must keep a-movin' it. Leave it be in th' wet an' it'll goo black an' mouldy, then you got ter burn un. An' a bit o' poor 'ay's better'n no 'ay at all."

"That's right, Fred," said the boss, "and the amount of time you spend on it, poor hay's aways a bloody sight more expensive than good, quickly-made hay."

The time came when the hay rustled and rattled crisply when we stirred it with a prong, and after much twisting up of handfuls the boss and Fred decided that the stems were no longer sappy enough to cause a risk of overheating and spontaneous combustion in the stack.

"That be just about fit." said Fred. "Dry it a touch more an' the leaf'll get too brittle an' drop off – then us'll 'ave lost the best part of the 'ay."

Instead of spreading the rows that morning Fred went round with the side-rake, first clockwise and then anti-clockwise, turning two, two-swath 'nightrows' into one big 'windrow' of four swaths, each eighteen feet apart. We went round with prongs once more, fluffing the hay up and teasing out the few remaining pockets of damp stuff, then the field was ready to be carted and stacked.

In previous years hay had always been pitched onto wagons in the field, carted to the side of the stack and then pitched off again. But this year the boss decided that we would go modern and use the Ferguson and buckrake. The new tractor came into its own once more, speeding backwards to the stack, a small mountain of fluffy hay pushed along by the buckrake. There were one or two heart-stopping moments when the tractor's underslung exhaust blew sparks into the hay just behind the back wheel, causing a hasty leap off the seat and a bit of furious stamping-out, but luckily it never happened near the stack. Everyone said how much better the new method of carting was except, of course, Fred, who grumbled.

"When yer pitches on an' off a wagon there's nice pitches o' 'ay what builds in the stack a bit sensible, like. That bliddy thing do roll an' twangle it up all ways. One minute I gets no mor'n a 'atful pitched up ter me an' the nex' thing's a bliddy gurt solid lump weighin' a 'undredweight. Yer can't build a stack no sense like that. Bliddy thing'll fall over, you see if it don't."

But despite his predictions the stack grew up as straight and true as ever, with the middle well filled, ready to withstand the worst of winter's storms.

During the winter months the boss was a keen rugger player with the Dorking RFC, but it was in the middle of the summer during haymaking, that Fred and I were given a demonstration of the turn of speed that had earned him a place in the club's 1st Fifteen.

We were taking the sheet off a half-completed rick one morning, ready to start the day's stacking, and the boss was walking towards us across the field. Fred suddenly stopped folding the sheet, pointed and said, "What the 'ell's boss playin' at, then?"

I looked, and there was the boss, crouched in the middle

of the field. He waved his arms over his head, then sprang to his feet, sprinting round in circles and figures of eight. Then he halted, crouched again and stared upwards. Fred looked at me and shook his head sadly.

"I reckon 'e's gone wrong in the 'ead, Boy." He leant on a prong handle, peering across the field, his jaws working rhythmically on his chew of tobacco. "They gets like that durin' 'aymakin', some on 'em."

In a final burst the boss darted straight towards us and fetched up against the side of the rick, panting and gasping. Fred peered over the edge at him.

"'Ullo, Boss. 'Ave a good run, did yer?" The boss glared at him, unable to speak for a moment, he was panting so hard, then he gasped out,

"Didn't you see 'em?"

"See what, Boss? All we see was you runnin' like you was gone daft."

"Bloody bees," said the boss angrily. "you must've seen 'em."

Fred and I looked at each other.

"Bees?" I said.

"Damn great swarm of 'em – millions of the buggers. Walked right into the middle of 'em I did."

"Ar, bees, was it?" said Fred. "Yer wants ter steer clear o' them things, Boss. They can be real nasty little devils, bees can" he added helpfully, "Stings yer, they does."

Fred was very quiet and thoughtful for the rest of the morning and only spoke when we were walking home at mid-day.

"Do yer 'spose 'e's all right, the boss, I mean?"

"I don't think he got stung," I replied.

"We'd have heard all about it if he had been."

"I don't mean stung, like..." Fred shook his head doubtfully "...I don't mean that. An' I knows as 'ow I be a bit short-sighted, like, what wi' me eyes not bein' what they was – an' I dunno whether you 'see em, Boy – but I never see no bliddy swarm o' bees!"

For the next couple of days, while we finished off the haymaking, I noticed Fred eyeing the boss furtively from time to time. But as his behaviour appeared to be normal – or as normal as any farmer's is during haymaking – Fred must have given him the benefit of the doubt, deciding that his sanity – and our jobs – were safe for a little while longer.

8

AUGUST

Fred was in one of his rare grumbling moods as he and I got the old Albion binder out of its winter quarters behind the stable. It was the beginning of August and Fred and the boss had fallen out over whether we should bother to cut round the headlands of the corn-fields with a scythe before the binder started cutting. The object of 'opening-out' the corn was to cut a three-yard passage right round the outside so that the horses – or nowadays the tractor – that drew the binder wouldn't trample the standing corn into the ground. Fred had already opened out the little four-acre field next to the railway and I had been sent with him to tie the mown corn into sheaves so they could be stacked with the rest of the crop when it was cut with the binder. Fred was a pretty fair hand with a scythe but, even so, it had taken him six or seven hours to cut round the four-acre field. The boss had worked out that there was at least another week's work if we opened out the rest of our forty acres of corn, and there simply wasn't time to do it.

"It's not necessary anyway, Fred." he said as we discussed the day's work first thing that morning. "The little bit of corn you run down with the tractor wheels gets picked up as you go round the other way, like you do with a grass-mower, taking out the back cut."

Fred carried on grumbling after the boss had gone off to stir someone else up. He enjoyed jobs like mowing with a scythe, hedge-laying, rick-building and thatching, because he was the only one on the farm who was a real expert at them – indeed, at these now out-dated crafts Fred had no equal in the district. But his wasn't a selfish 'look how clever I am' attitude, it was the real joy and satisfaction of being a master craftsman. Being deprived of the chance to practise the skills acquired in sixty years on the land – and to pass them on generously to people like myself, the next generation – upset him and made him crotchety.

"Tis all a-rushin' an' a tearin' about these days. There an't no time ter do a decent job o' work like us used ter."

I knew my place better than to join in the argument. I had enjoyed tying up the sheaves with Fred – tying a bond made of the stems of corn round them as had been done since farming first began – as it made me realise the enormous amount of labour involved in the harvest before even simple machines like the first mechanical reapers made the work ten times easier. And another reason for my silence was that I was secretly enjoying Fred's dispute with the boss. The two of them had given me a hard time the day before over the matter of the rick iron.

The last hayrick we put up had been a rush job. We wanted to get the haymaking finished so that the corn harvest could start, and the last piece we carried could have done with another couple of days drying. But the boss decided to carry it 'a bit on the gay side'. Surprisingly, Fred hadn't objected to this and had taken it in his stride.

"Tis a bit on th' greenish side, Boss, but that'll be all right if us puts a chimbley in th' rick as we builds un."

Making a chimney in the rick was done by drawing a

large faggot of sticks, or a truss of straw, up through the centre as each succeeding layer was built on. When the peak of the roof was reached the faggot was taken out and the chimney – a duct about two feet across – remained open from top to bottom of the rick. Then if any undue heat was generated in the centre, it was vented out through the chimney before the temperature could build up to the dangerous level where spontaneous combustion occurs.

Our rick had been built with a 'chimbley', but it had started to get quite hot. The boss was worried, and there had been an earnest discussion at the rick when we all thrust our arms as far as we could reach into the new hay, withdrew a sample and sniffed at it knowingly.

Ernie and Fred said that it was as safe as houses, but they didn't have the ultimate responsibility. The boss, mindful of the fact that a hayrick wasn't covered by insurance until three months had elapsed – time for the risk of fire from natural heating to have passed – scratched his head and said he wasn't so sure.

"That be all right, Boss," said Ernie as we walked down the lane to the farm. "it's getting' warm 'cause it's a good bit o' stuff – good clover there is in that rick."

"Ar, that's quite right." said Fred. "'Ay what don't warm up a bit in th' rick han't got no guts to it." He paused to squirt the juice from his chew of tobacco into the hedge and added, "An I'll tell yer another thing – that's the best 'ay fer 'orses, when it been burned a bit brown in th' rick. Loves it, they does." The boss shook his head doubtfully.

"I just think it might be getting a mite too warm, that's all." Ernie tipped Fred's cap playfully over his eyes.

"Well you knows what ter do, old feller, don't yer? Put the rick iron in un."

I pricked up my ears. I hadn't heard of a rick-iron before, but was wary of asking what it was. Sometimes, when I displayed my ignorance, I got teased unmercifully – it was all part of the process of turning pompous sixteen- year-olds into competent farm workers – and I could see that the three men were in one of their boy-baiting moods.

"Please," I asked timidly, "what's a rick-iron?"

They stopped as if they had run into a wall and looked at each other wonderingly. Then all three of them stared at me in exaggerated horror. Ernie broke the silence.

"I'm real surprised at you, Fred," he said. "You've 'ad boy workin' along o' you fer nigh on eight month now, an you han't learned 'im better'n that?"

" 'Taint my fault, Ern." Fred looked suitably doleful. "I've a-tried ter show un proper, like, but what can you do? Shows me up summat terrible, 'e do, sometimes."

The boss clapped his hand tragically to his brow.

"We've all failed," he said brokenly, "we're all to blame. It's not Chris's fault that he's as thick as three short planks. He must have fallen over the rick-iron almost every day in the barn – one of us should've told him what he was bumping into."

"That's right, Boss." said Fred. "Us's done wrong by un. But us better put it right a bit sharpish or e'll grow up ig'orant like one o' they fellers from the Ministry." He turned to me.

"Yer know when yer mum presses yer trousis of an evenin', afore yer goes out? She uses a flat-iron, don't she? A rick-iron be summat like that on'y a bliddy sight bigger."

"Much bigger." said the boss. "The one we've got in the barn must go nearly two hundredweight. And it's a small one."

"Two hundredweight?" I said doubtfully. "What's it like?" Ernie nodded wisely.

"More like two an 'a 'alf. An' I've seen bigger ones nor that, I 'ave."

"That's right, that is." said Fred. "When I were a lad down Sussex us 'ad rick-irons what took four 'osses ter drag 'em to th' rick." The boss smiled reminiscently.

"They use camels in North Africa. I saw 'em out there during the war. Great long string of camels all harnessed together. And sheer-legs and pulleys to get the iron onto the rick. Of course," he added, they don't make their ricks like we make ours – theirs are shaped like pyramids."

"Pyramids?" I said. "Up on the rick? I don't believe you." Fred snorted.

"Them as doesn't b'lieve doesn't learn nothin'. O' course yer got 'em on the rick. Yer rolled that silage 'eap wi' the Fordson, didn't yer, ter cool un down? Well yer can't get no Fordson on no 'ayrick so yer uses a big 'eavy rick-iron instead."

They were still telling me about bigger and more complicated rick-irons they had used when we reached the yard. It was a long, thin rod with a loop handle at one end and a small, barbed point at the other. It was used by being thrust into the rick so that the point was in the centre, and left overnight. In the morning, when it was pulled out, the temperature of the metal was the same as that of the rick, and the barb withdrew a small sample of hay from the middle, so that it could be seen if it was heating and going brown.

Our hayrick, when we tested it, proved to be at a low, safe temperature, much to the boss's relief and Fred and Ernie's satisfaction.

But that was yesterday. Today we were getting the binder out. Its proper name was reaper-and-binder, in the same way that a combine should properly be addressed as 'combined harvester/thresher' and it was the end of an evolutionary chain in the development of harvesting machines; a chain that started when the first Neolithic farmer put sharp flints into the curved jawbone of a deer and made a primitive sickle, to cut his thin crops of wild cereals instead of plucking them.

After 8,000 years of development, through the stages of the sickle, reap-hook, bagging-hook, scythe, cutterbar mower and reaper – the binder was the ultimate machine. It was a totally new concept as it did all the jobs connected with reaping in one operation. Cutting the corn, gathering it up into bundles of exactly the right amount for one sheaf, and then tying a string round the sheaf before throwing it out on the ground ready for stooking. It sounds simple but it was a very advanced idea in terms of the automated handling and packaging of materials – and a product of the inventive Victorian age. The combine was also a product of the same age (the first one was made in the late 1830s – in America, of course) but it wasn't a new concept. The combine was a marriage of two existing machines, the mechanical reaper and the threshing machine.

However, we weren't concerned with combines ('New-fangled, gurt old things they be, they won't never catch on hereabouts – our fields is too small for 'em'), and Fred cheered up as we dragged the rusty old machine out of the lean-to and cleaned the muck and cobwebs off it. There is always a glad and excited feeling of anticipation about the start of the harvest – the culmination of the farming year – and even an old man who has seen it all before about sixty times and who is still suffering from the effects of being bested in an argument, couldn't help being uplifted by it.

I watched Fred as he went round with an oil can, filling up oil cups and squirting oil into the holes above each of the many bearings.

"Aren't there any grease nipples on it, Fred? I thought bearings had to be greased."

"Lor, bless yer, no, Boy. This 'ere 'chine were made long afore they thought o' things like that. Been on this farm longer'n I 'ave, this old binder 'ave."

And it looked it. There wasn't a trace of paint anywhere on the complicated metal frame, and the wooden parts – except where they were stained black by over-liberal oiling of adjacent chains and bearings – were baked a rich brown by the hot sun of many harvests.

We got the slatted canvas conveyor-belts – the 'canvasses' – out of the shed where they had been hanging on strings under a beam, out of reach of rats and mice, and fitted them in place. I was then able to work out how the corn was cut by the knife, tipped back onto the moving canvas behind the cutterbar by the sails, and carried to the centre of the machine. There it was picked up by two more canvasses and went up between them to fall onto the 'deck' where the packer arms formed the stalks into the shape of a sheaf, up against the trip lever of the knotter. When the sheaf was large enough, its weight pushed the trip down, a clutch engaged, the curved twine-needle carried the twine around the sheaf to the knotter, where the bill-hook – a double finger shaped like a bird's beak, hence its name – opened, grasped the twine, tied it into a knot and cut the ends off neatly with a sharp blade. Then three arms swung over, kicked the complete sheaf off the deck and the knotter came to rest, ready to be startled into its next jerky cycle by the pressure of another sheaf.

"Will it work Fred?"

"Allus 'as done. Bliddy good binder this be. Well, should be an' all – 'er's 'ad plenty o' practice."

Farmworkers always speak kindly to their machines and praise them as well as lubricating them before a job starts. It's well known that each machine has a mind of its own, and the more complicated they are, the more cantankerous they can get. Part of the art of farming is to know just how much, and when, you have to be nice to an individual machine. Insufficient praise and the machine will refuse even to start, and will lie there in sullen and rusty rebellion. And if you overdo it with gratuitous flattery at the wrong time, the thing will hear you and get big headed – then a major calamity will occur just to spite you.

In my inexperience I breached the rules of etiquette that first day out with the binder. We'd stopped to make an adjustment with only a small piece of corn left to cut. I remarked on how wonderfully well the binder went considering its age. The boss looked horrified, grabbed my arm and actually pulled me away from the machine so that it wouldn't hear what he was saying.

"For God's sake don't say things like that," he hissed, "it'll break down, sure as a gun."

"I'm sorry, I said, startled, "I was only saying how…"

"Don't do it again," he said, "it's bloody unlucky."

Then he went back to the binder and surreptitiously touched the wood of one of the sails. He saw that I'd seen him do it, blushed red, got on the tractor and the binder rattled off, throwing out sheaves in fine style. It didn't break down that day so the wood-touching must have done the trick – or maybe the binder forgave me, knowing that I was young and ignorant.

Fred sat on the seat at the back of the binder with all the controls near to his hand. The only things missing were the reins – the iron loops deigned to keep the reins clear of the mechanism of the machine were empty – as the tractor was towed by a tractor, but I noticed a stout stick in the rusty whip socket. I couldn't think what it was for, until the knotter jammed and the binder started throwing out sheaves with no string round them. Then Fred beat lustily on the metal casting with his stick and the boss heard the banging and stopped the tractor while Fred cleared the knotter.

If Fred had merely shouted, his reedy voice would have been drowned by the roar of the Fordson's exhaust and the grind of its transmission which could be heard a mile away.

When a third of the field had been cut, they stopped for a break and the boss gave me a lesson in stooking. I had made one or two stooks while the binder went round and thought it was an easy job, but Fred had been eyeing my efforts critically as he passed by on the binder. I felt a little hurt at this. The stooks I'd made didn't look very tidy, I admit, but the sheaves were the right way up and there were six of them to each stook. What more could you want?

"Yer reckons that be good enough, do yer?"

Fred walked across to the offending lump and gave it a gentle push. It fell over and the sheaves splayed apart.

"An' that weren't more'n a little old puff o' wind. Ask the boss ter show yer 'ow ter stook proper."

Fred walked back to oil the binder and the boss showed me that the butts of the sheaves were angled slightly from the action of the packers. If two were put with their heads shaken together, the right way round with the knots of the string outwards, the butts apart so that the sheaves formed an inverted 'V', they leaned naturally against each other, giving mutual support, with the butts flat on the ground. In the same way, another two pairs of sheaves were placed on either side of the first pair, leaning slightly inwards. The whole stook of six then stood as steady as a rock, ready to withstand a gale of wind.

" Make sure you stand 'em good and firm. Any that fall over have to be put up again and it's a waste of time to do the job twice. Remember they've got to stand for a while, too – they have to hear the church bells three times before we cart them."

The boss meant that the stooks would have to stand for at least a fortnight – three Sundays – before the ears had ripened off and any greenstuff in the butts had wilted and dried.

"Cart 'em before they're fit and you'll get mouldy corn and a hot rick. But there's no risk of that happening on this farm – Fred wouldn't let us do anything so stupid."

The self-appointed master of the harvest field had finished oiling the binder, had filled and lit his pipe, and was sitting impatiently back on the seat on the binder, so the boss had to hurry back to the tractor. But before he went, the boss ignored the pointed mutterings, 'Time be a-goin' on!', for a few seconds while he gave me another tip. This was to put my shirt back on – I had thought real farm-workers should look bronzed and sunburnt – and roll the sleeves down, buttoning the cuffs. If I didn't, he said, the insides of my forearms and any exposed skin would get red raw in a short time through handling the bristly sheaves.

As the afternoon wore on I got more proficient at stooking and took a great pride in my work. There was a feeling of achievement, looking back and seeing the lines of stooks standing like soldiers on parade. Each line the correct distance from the next, so that the harvest carts could load two lines at once, and each stook opposite its neighbour. Then the wagon could halt between four stooks and be loaded with the minimum number of moves.

The clean-cut, short stubble and the regiment of sheaves in their platoons of stooks were pleasing and orderly – much more so than the shaggy cut and untidy heaps of straw left by a combine.

The binder clattered on round the field, throwing out sheaves faster than I could keep up with them with my stooking, and the square of uncut corn in the centre of the field grew smaller and smaller. Rabbits that had grown up and spent their short lifetimes amid the waving stalks found their whole world being destroyed by the noisy machine. They retreated into the dwindling refuge until that became too over-crowded, when they were forced to make a run for freedom across the unfamiliar and exposed stubble. They were dealt with by the boys of the village who 'happened by' when they heard the binder running. I knew all about this 'happening by' as, in the far-distant days of my youth a whole year previously, I had taken part in it.

Every village had its gang of boys and the Abinger one was quite the usual thing. It was ruled dictatorially by the leaders of the strict hierarchy, and had a rigid code of conduct and behaviour, with penalties for infringement ranging from a mild arm-twisting to the ultimate horror of banishment from membership forever – a savage sentence for the most serious crimes like 'telling', associating with girls or a rival gang, or, worst of all, failing to attempt a dare through cowardice. Banishment 'forever' sometimes lasted for up to a week – an enormous length of time and the limits of human endurance to boys who are by nature gregarious animals. I missed most of the delights of growing up in a village by being sent to a boarding school at an early age, and consequently was never allowed full membership of the Abinger gang. I was, however, allowed a form of associate membership during the school holidays provided that I behaved myself as a visitor should, and never made any attempt to have any say in the running of things.

The meeting place for the boys was the bridge over the stream opposite the post office. We used to gather there and lean on the railings, spitting into the water and watching the brown trout and gudgeon, while the day's mis-deeds were planned. But in the days of early harvest all attention was on listening for the distinctive drone of a Fordson pulling a binder on one of the five farms surrounding the village

"My dad says they're cuttin' at Blackhurst today."

"Can't go up there 'cause 'e's a rotten ol' bugger, says we gets in the way of the guns. We'll wait 'til Newman's nearly finished that barley by Piney Copse. I can 'ear 'im goin' now. Right then – bet yer couldn't jump across to that bank there 'thout fallin' in the water – I could."

Some farmers organised men with guns to shoot the rabbits as they ran out of the corn, but Bob Newman was one of the enlightened ones who realised that four or five people on each side of the square was a much more effective way of dealing with them. And a lot cheaper than cartridges at nearly a shilling each.

When the gang of boys appeared in the field we were cutting I remained aloof and ignored them for a while, continuing with my stooking as if they weren't there. The leader deployed them around the field with the eldest and fastest runners nearest to the railway embankment where the rabbits were most likely to run, and then took up a position near where I was working. I acknowledged his presence with a curt "Keep clear of the binder" and he replied with a respectful nod. Then we grinned at each other, having established that I had crossed the vast gulf between boyhood and working manhood. It was strange – only a year previously, although he was younger than I, he had been the leader and had been bossing me about with calm authority. Now our roles were reversed.

It wasn't only boys who came to join in the fun. There were men, too – shift workers like the postman and policeman who had time to spare, and others who hadn't but dropped their tools on the railway line or in gardens in preference for the excitement of the chase.

The rabbits were easy to run down and kill. They stumbled aimlessly over the prickly stubble and cowered under sheaves and stooks. We didn't stop to pity them – they were a serious pest in the days before myxomatosis, to be despatched with a quick bang on the head. And they made a tasty addition to the small meat rations we got from Reads, the village butcher.

When the last, small strip of corn had fallen to the knife of the binder, and when the last rabbit had run, we gathered round and counted up our spoils. The dead rabbits were shared out, one or sometimes two to each person who had joined in the hunt, with a couple put by for Ernie who had been busy milking. Then the boss collected up the remainder. They were for people in the village who had done him a favour, or sometimes for pensioners too old and feeble to come out into the harvest fields and catch their own rabbit pies.

The boss and Fred packed up the binder ready for the next field, and the rest of us got busy and finished off the stooking. It didn't take long with so many helpers and it wouldn't have done to have left sheaves lying on the ground overnight.

The railway gangers went back to their shovels and crowbars on the line, on which they leant in rock-like immobility whenever a train passed by; the postman returned to whichever of his part-time jobs he was supposed to be doing that afternoon – he was chimney-sweep, barber, jobbing-gardener and general factotum in the village – and the policeman reverted to his normal, dignified self. Our respect for the Arm of the Law hadn't been diminished at all by the brief glimpse we had been given of the human being under the uniform – whooping and yelling and stick-waving with the best of us. He bent, with a creaking of braces, to fasten his bicycle clips then mounted his tall bicycle. Two rabbits hung from the handlebars.

"See you tomorrow, Mr Bishop? We're cutting the Six-Acres by Hammerfield."

"If my duties permits, Lad."

9

SEPTEMBER

The bustle of harvest gathered momentum and days off and weekends were ignored in the rush to get the corn safely into the protection of ricks. Days were long, and seemed to merge into one another so that August slipped away unnoticed. Before we knew where we were it was September and with the binder safely put away in its shed for another year, carting and stacking got into full swing.

Each morning Fred and I went off early to the cornfields to get the preparatory jobs done, while the boss got the harvest gang organised – jobs like making the staddles, or bases, of the ricks out of faggots and trusses of straw, and knocking down enough stooks for the day's work so that the butts of the sheaves would have dried out by the time they were pitched onto wagons.

Of course, the weather was the governing factor in deciding how many stooks to knock down, or whether it was safer to leave them as they were to shed the impending rain, with the half-completed rick sheeted and covered. For this reason Fred's first task, first thing each morning, was to give the boss a detailed forecast on the prospects for the day. He was seldom wrong and used a variety of signs in coming to a decision regarding the prospects for the next twelve hours or so. The main one was the direction of the wind and this was indicated by whether clocks could be heard striking from Shere, to the west; from Abinger Common or Peaslake to the south; or from that infallibly dry quarter, the stable clock at Abinger Hall – due east of the village. Other signs were the redness of the sunset the night before, the state of the sun when it rose at dawn, the amount of dew on the grass (a heavy dew and a clear sky means a fine day to come for sure), and the casting vote was always held by Fred's bones. "Me bones be achin' – 'tis goin' ter rain afore night!'

The boss used to listen to the BBC's forecasts on the radio but Fred was very scornful of these.

"'Ow do they wireless chaps know what's a-goin' ter 'appen down 'ere? Why, it could be pissin' down 'ereabouts an' they wouldn't know nothin about it – 'tis bliddy miles away that Air Min'stry roof o' their'n, up Lunnon way."

Fred had me fooled one day when we were haymaking back in July. It was one of those blistering hot days with no wind, the North Downs shimmering blue and distant – they look close enough to touch when there's rain about – and not a trace of a cloud. Fred suddenly stopped, stuck his prong in the ground, spat some tobacco juice and said,

"Tis a-goin' ter storm wi' rain, Boy!" I was amazed at this and thought I was going to learn yet another secret sign of rain that the old man hadn't told me about. I said,

"How can you tell, Fred?"

Fred looked at the sky, looked all round the horizon and then looked back at me in a thoughtful way. Then, without another word, he picked up his prong and continued working. I asked him again,

"How can you tell?" He grinned at me and said,

" 'Eard it on the bliddy wireless 'smornin'."

Fred carried his tools with him – a faghook to cut nettles away frow the site of the rick, and his short-handled prong. The prong was his badge of office as the most important person on the farm during harvest – a badge more significant to a countryman than any mayoral chain is to a town – the hard-earned sign of the master rick-builder. To him was entrusted the safety and well-being of the whole year's work. Badly made ricks would let in the wet and there would be losses and a poor-quality sample of low-priced grain to offer the corn merchant. And in these days the corn merchant acted as the farmer's banker, extending him credit throughout the year for the purchase of seeds, fertilizers and feedingstuffs. The debt was repaid by the farmer when the corn merchant bought the farmer's grain from him when it came off the threshing machine in the autumn or winter – so a poor harvest, or grain spoilt in badly made ricks could spell financial disaster.

Not that Fred showed any sign of his burden of responsibility as he walked up the lane to the field. The early morning mist rose off the Downs ahead of us and the cool air was newly washed and scented – acrid smoke from Fred's pipe making a sharp contrast as he puffed contentedly.

"Goin' ter be another scorcher today, Boy, there an't no wind. Yer can 'ear old Ernie milkin' a mile away."

That was the new sound in the village – the mutter of the little petrol engine and the whine of the vacuum pump as Ernie milked our increasing herd of cows by machine on the portable bail. He was alone with his cows on the top of the Hangers the other side of the valley, but the sounds carried well in the still air. We could hear him singing as he worked; the clatter and ring of buckets and milk churns, and occasionally – if one of the cows misbehaved – an outburst of bad language and recrimination. Ernie seldom hit his animals, and if he did, it was with the flat of his hand which didn't hurt their thick hides in the slightest. But to see one of the big animals cowering away from Ernie when he was displeased, you would think he was a real ogre – until you saw the twinkle in his eye.

Ernie's job depended on a proper and genial relationship between himself and his cows. A cow can't be forced to enter a portable bail in the middle of a field – she has to be made to go in of her own accord. And sometimes – cows being the cantankerous creatures they are – they have to be given the impression that the *last* thing that is wanted is for them to come in and be milked. Then the awkward one will push past the others into the bail just to spite the cowman – or so she thinks.

Ernie was the master animal psychologist on the farm just as Fred was the master rick-builder.

To get the harvest in before the bad weather of autumn arrived needed a lot of organisation and a lot of manpower. It was usual to have at least two wagons picking up the sheaves, so that one could be loading in the field while the other's load was being pitched off at the rick. Each wagon needed two pitchers and one man to build the load, and two more were needed on the rick, one to do the actual building (Fred, of course), and the other to pitch sheaves to him. So the minimum requirement was eight people – with a further saving in time if there was someone on the tractor to move it between the stooks in the field. Theoretically it shouldn't have been necessary to have a driver for Kitty, the mare, she should have stopped and started at the word of command. She'd start all right, but she'd swerve off straight to the nearest stook and start gulping down the ears of corn. She had to be led firmly down the centre of the row and be told to 'Stand, yer owd booger', which was Fred's customary order to her which she understood and obeyed.

The gang of harvesters provided to supply the 'casual labour' were friends and relations of the boss and his wife. They had been lured onto the farm from their suburban jobs and homes with tales of 'a nice holiday on the farm'. They stayed in the farmhouse, were fed enormously by the boss's wife (who consequently had to work twice as hard as anyone else during harvest – and in a hot kitchen, too), and really earned their keep. They sweated and strained, making the job of pitching sheaves ten times more difficult because they didn't have the knack of it, but supplying the necessary manpower. Manpower needed because each ear of corn, from the time it stood ready for cutting in the field, to the time when the stream of grain poured out of the thresher into the two-hundredweight sacks, had to be handled at least twelve times.

Fred took charge of the gang with an easy authority, while still maintaining the pretence – in the subtle way that key men in farming have – that it was the boss who was directing operations. We were pleased to have the gang on the farm as they did most of the donkey work, but Fred and Ernie were a little scornful of their amateur status – amateur on the farm, that is. They were mainly professional people with incomes far in excess of Fred's wildest dreams of wealth; if ever he had dreams of that sort, which he probably didn't. He had his house and garden rent-free, a big allotment in the village for five shillings a year rent, the village pub for relaxation and a bit of gossip, and his work on the farm which was his all-absorbing interest. He was quite content with this and happy with his station in life which, as he "never 'adn't 'ad no schoolin'", meant deferring to those with the benefits of an education. He touched his cap to most people he met in life – not in a servile way (Fred was anything but servile) – but because it was right and proper and good manners to do so.

But up on a corn stack Fred touched his cap to no-one. He was the natural leader, the master craftsman at work. Even the boss deferred to him then and you offended him at your peril.

Fred's first job was to mark out the staddle for the rick, exactly the right size for the amount of corn that was to go into it. And the site of the staddle had to be chosen with great care. It had to be on rising ground so the run-off from the eaves would drain away, rather than into the base when the rains of winter set in; space had to be allowed for the thresher and baler, and the thrashing contractor's big tractor which drove the whole outfit with a long endless belt, when the rick was threshed out some time near Christmas; and there had to be a suitable site left for the straw stack, remembering that it must be near where the baler or trusser would stand. Another factor in the calculation was that the straw stack had to be easily accessible during the wet time after Christmas – it's no good having a stack of good straw in the field when your wagons and carts bog down as soon as they're loaded.

I don't think Fred ever thought of all these things when he was deciding where to place a rick – ticking them off on the list in his mind, as it were – it was just the weight of experience bringing the right answer in the short time it took to fill and light a clay pipe. Then it was,

"Us'll put un 'ere, Boy. Lay them faggots down good an' tight together where I be a-puttin' these pegs in."

It was interesting that corn was always built into a 'rick', while straw was a 'stack'. At first I thought that the two words differentiated between round and oblong ricks or stacks, but the locals didn't seem to think this was the case. 'Never makes round ricks round 'ereabouts anyways, Boy, they'm bliddy awk'ard ter get right.' So it must have been local usage and ease of pronunciation in our county's dialect. Certainly, 'cornstacks' and 'straw-rick' don't trip off the tongue as readily as 'cornrick' and 'strawstack', whereas hay could be made into either. 'Haystack' and 'hayrick' are easy to say.

Of course, someone writing about another county – or even another part of the same county – would use the words the other way round and be correct for his locality. The heavy wooden mallet used for driving in fence posts is described in the dictionary as a 'beetle'. In Kent it's called 'beadle', in Sussex a 'bittle', in Hampshire a 'bightle' (rhyming with 'light'), and in Wiltshire a 'boytle'. If you asked today's farmworker – with lots of GCEs and a degree or two – what he called the implement he'd probably say, 'Never use one, old boy, not cost-effective. Use a tractor-mounted hydraulic rammer.'

The rick builder always went out with the first wagon and decided which part of the field would be carted first. The best and driest sheaves had to go into the bottom of the rick so that there was no possibility of heat being generated which would discolour the grain. Any that were slightly damp or had a bit of green in the butts were left until last so that they went into the roof where they received the maximum ventilation.

I was shown how to load the old Sussex 'boat wagon' – so-called because its body was shaped like a hollow hull – and given a few tips on how to control my gang of pitchers.

"Fill the well o' the wagon first, then lay yer sheaves round like I showed yer, butts out'ards. Keep yer middle up all the time so's the outside uns is pinned-in an' won't slip. Fill in good an tight when yer be toppin'-out, an' make bliddy sure they shouts "Old Tight!' afore movin' on. That old mare do throw 'erself into 'er collar ter get a 'eavy load movin' an yer'll tumble off the load if yer an't 'spectin' it. An' if they pitches up too fast fer yer load proper, push 'em off t'other side. They'll bliddy soon get fed up wi' pitchin' 'em up twice."

And he delighted me by saying to the ex-army major and the well-to-do estate agent who were going to pitch to me,

"Go along o' young Chris, 'ere, an' do just as 'e shows yer. 'E'll put yer right."

They knew that I was a green youngster but Fred's words had been said to give me confidence, and his use of my christian name instead of the usual 'Boy' indicated that he and I were professionals – 'permanent staff' – and that I carried some of Fred's vast authority when he placed me in command. Nobody, not even an ex-army major, would have dared challenge Fred's chain of command.

The rick rose steadily and squarely as the expert worked his way round each 'laying', moving anti-clockwise on one, clockwise with the next. Each sheaf went into precisely the right place, the right way round and the right way up – butts sloping down towards the outside of the rick. Then there was no possibility of water running into the rick, it was shed off as it is with a thatched roof. And the area of the rick at the eaves was greater than the size of the base so the sides sloped outwards, adding further protection against wind-driven rain.

The person pitching sheaves across the rick to Fred had to deliver them the right way round direct to his hand. He wouldn't bother to reach for them and would ignore any that were badly placed. And if anyone dared to lay a sheaf themselves, that caused an immediate upset. He'd glare at the offending sheaf, pick it up and re-lay it properly, sometimes in exactly the same place. Then he'd spit contemptuously.

"I be makin' this 'ere bliddy rick!"

We used to swear that the ricks Fred made couldn't possibly fall down. He couldn't smoke on the rick but he had to have tobacco, so he'd chew all day long, his jaws moving rhythmically on black shag from the Co-Op, the same as he smoked in his pipe. Each sheaf was treated to a squirt of tobacco juice as it was laid – or as we said, glued into position. Certainly, our ricks seldom needed propping at the corners, or where a side had started to lean, by ramming a stout pole into the side like a buttress, taking the weight and preventing further slipping.

I got into terrible trouble one day when I forgot to tie the mare's head away from the rick. I had just brought a load in, and while I pitched the sheaves off, Kitty reached her blinkered head round and with wide-flapping lips pulled a couple of sheaves out of the corner. The corner started to slip out. Fred was furious.

"Look what yer bliddy well done, Boy! Us can't get that back again, not no 'ow. We'll 'ave ter put a prop agin un now. 'Tis damn lucky fer you, Boy, as 'tis the corner away from the gateway."

That was what really upset Fred. One of *his* ricks had a prop against it. All the neighbours would walk up the lane, see the prop and comment on it. And as Fred was acknowledged to be the king of rick-builders in the village, the thought of his shame – or rather, the shame I had brought upon him – was almost too much to bear.

It happened that the only person who dared to comment on our propped rick was a notoriously bad rick-builder from a neighbouring farm. He sneered at Fred about it when we were in the Abinger Arms one evening after harvest. At first, I thought Fred hadn't heard the taunt, but he was treating it with the contempt it deserved. He took a long pull at his pint, wiped his mouth with the back of his hand, turned to me and said, "If George be on about proppin' ricks, Boy, I'll tell you summat. I been round an' seen' some o' them lumps what George built. Most on 'em 'ad props an' one 'ad seven or eight." He took another mouthful of beer and his eyes twinkled. "Why, Boy, that 'ad so many legs it could've up an' walked around the farm like a bliddy gurt beetle."

George didn't say another word. He retired to the other end of the bar and ignored us for the rest of the evening.

"Take a lesson from 'im." said Fred. "Don't never start shoutin' 'til you'm sure yer knows better'n t'other chap." He emptied his glass and wiped his mouth once more. "An' even then," he added, "'tis better ter keep yer trap shut." The empty glass banged down on the table between us.

"Now then, Boy, I never told George as 'twere your fault us 'ad ter prop that rick. I reckon you owes I a pint."

I bought him one.

Each evening, at about six o' clock, the boss's wife and the cowman came up to the field with baskets of tea, sandwiches and cake. That was the time of day when tiredness and the heat of the day were at their maximum. Long hours of sunshine had dried us out to the point of dehydration, so a half-hour break was necessary – it enabled weary bodies to get their second wind.

China mugs were filled and handed round, and we sprawled with our backs against the wheels of wagons, teasing Ernie about his fresh, clean appearance.

"Easy ter see who's been doin' all the work today, an't it? Some on us look like they han't done a 'and's turn all day." Ernie grinned good-naturedly.

"I done my share today, I 'ave, wi' them old cows. It were that sweaty 'ot on that old bails' afternoon, and what wi' the flies an' all, I reckon you got the easiest job up 'ere. So I 'ad a bit of a wash an' brush-up an' changed me shirt in the dairy afore I come ter show yer 'ow to do a bit o' graft."

Fred spoke through a mouthful of cake.

"Ar, I were just admirin' that nice white shirt you got on. Looks a real gent, yer does." He waved his mug in the direction of the boss, the land agent and the major, who were deep in earnest discussion about land prices, the iniquities of Mr Attlee's government, and the outbreak of war in Korea. "Reckon yer ought ter be a-sittin wi' that lot, 'stead o' us workin' chaps."

"I think I'll be a cowman later on, Fred." I said. "With all the money they earn they can afford to be snappy dressers like Ernie. Think of all the girls I could get if I went around in shirts like that."

"Don't yer start that bliddy caper, Boy." said Fred. "Rollin' about wi' them gals is a tirin' old job – I knows – I done plenty of it in my time. Yer'll be up 'alf the night like an old tom-cat an' yer won't be fit fer nothin' in the mornin'. Yer goin' ter start that, yer wait 'til after 'arvest – we wants ter get a bit o' work out of yer fust. Aint that right, Ern?"

Ernie, who had four children, spoke with feeling. "You steer clear of 'em, Boy. You starts messin' about wi' girls an' afore you knows where you are, you'll be married wi' kids. Yer bliddy money don't go far then, I can tell yer."

Fred stared reflectively back down the years.

"When I was a lad, down Sussex, I 'member this gal what 'ad seventeen of us, one after t'other, fer a shillun a time. Seventeen bob she made that night." Ernie was shocked. "Cor, bliddy 'ell, Fred, don't talk like that front o' Boy! Give 'im wrong ideas, yer will." He thought for a moment then shook his head. "Seventeen! No – I wouldn't fancy that. I don't like goin' where someone else 'as trod – an' seventeen on 'em!"

"No more don't I, Ern." Said Fred. "that's why I made sure I were the first."

The talk, admittedly not of a very high moral or intellectual standard, went round in a circle until Fred, having had a quick smoke, knocked out his pipe and wandered over to the half-made rick. He raked loose straws down the sheer sides with the prong and batted the butts of any sheaves he thought stuck out a fraction, spoiling the perfection of his handiwork. It was a signal that the tea break was over. We handed our cups back to the boss's wife and she packed them away in the empty basket. "Thankee, Missus, that went down a real treat, that did"

The boss looked at Fred fiddling impatiently round the rick, groaned and got to his feet. "Proper old slave-driver, Fred is. Don't know why the hell I work for him. Doesn't give you a minute, does he? Now then – Chris, you pull the load into the rick and chuck it off. Ernie, you come with me for another. There's a prong behind the rick there…"

Soon we were back into the swing of stacking, grateful of the cooling evening air, and by the time I had pitched my load off the next one came rumbling up to the rick. The short tea break was forgotten as if it had never been and work went on until it was too dark to see. Fred climbed stiffly down – he needed a ladder now the rick was so high – and we left the last couple of loads on the wagon, ready to unload in the morning.

It was nice to get a complete field finished at the end of a day, but we always made a point of picking up all the sheaves that had been knocked out of the stook. Any that were left lying would have to be stooked up again so that they wouldn't get soaked by the heavy summer dew. The boss and Fred were such good estimators of the amount we would get through in a day that this didn't happen very often – only if there was some unforeseen mishap which held us up. Like the time we had the old-fashioned equivalent of a puncture on the old boat wagon.

I was just finishing the load – 'topping out' by filling the middle up well – when one of the pitchers called up to me,

"I say – is this wheel all right? It looks as if the tyre's coming off."

I climbed down and looked at the wooden wheel.

The iron tyre was very slack and looked as if another turn or two of the wheel would send it bowling off down the field on its own like a hoop. I didn't know what to do. I had watched the blacksmith and the wheelwright shrinking an iron tyre onto a wagon wheel – their workshops were side by side next to the post office – and couldn't imagine what could have gone wrong to make the tyre come loose. Just then the boss came by, driving the Fordson and the empty four-wheeled trolley up from the rick. He stopped and shouted, "Come on, don't stand about. Fred's waiting for that load."

I explained what had happened and the boss took the entire tool kit out of the Fordson's toolbox (a roll of baling wire and a rusty, 14" stillsons wrench), bashed the tyre back into place with the stillsons and wired it firmly back in three or four places around the rim. "There – that'll hold it until we can get it down to the river for a night or two."

I looked puzzled, and it was the major who came up with the answer. "Ah, I see what you mean, Bob. Like Constable's 'Hay Wain'. You stand the wagon in the water to wet the wheels."

"That's right, Stan," said the boss. "wooden wheels dry out in summer and the wood shrinks, letting the tyres go slack. You stand the wagon in the river and the wheels swell and tighten up. There's a saying, 'Keep the wheels wet and the body dry, and a wagon will last you a lifetime'."

I tried this piece of new-found knowledge on my mother when I got home that night.

"Do you know why they stand wagons in the river, Mum?" I asked as I watched her get my supper out of the oven.

"Mind out, the plate's hot," she said. "In the river? – yes, it's to keep the wheels tight. Everybody knows that."

The disappointment must have shown in my face because she added,

"Oh, I suppose it's all rubber tyres nowadays. They hadn't been invented when I was a girl."

We had to work extra late one night – until well after dark – and it was due to the boss disregarding Fred's advice. We had almost finished carting a field of wheat and the boss said, "How many more loads, Fred?"

"Two more easy 'uns," said Fred, but the boss decided that we would get it all on one big load to save time. Fred objected on the grounds that if the load fell off it would take three times as long to pitch it all on again. The boss said,

"Nonsense, it'll be alright."

"You be the boss, but I be a-tellin' yer – a load the size that booger got to be wun't ride to the rick."

I got on the wagon to load but Fred told me to get down.

"This 'ere load got ter be loaded a bit proper, an' you an't 'ad 'nough 'sperience yet. I'll load un. An' even then I don't 'spect 'er'll ride."

We just managed to get all the sheaves on. At the finish I could just reach the top of the load with a seven-foot pitching prong, and the others, who had ordinary prongs, were having to throw the sheaves at the top in the hope that Fred would catch them. Quite a few over-enthusiastic ones sailed right over the top and landed the other side of the wagon, accompanied by a muffed shout of 'Bliddy 'ell!' from above our heads.

The load was so high that Fred couldn't get down and said that he would stay on until we got to the rick, *if* we got there. He added that the load wouldn't ride. Slowly and carefully we drove to the rick, and all the way the voice from the top of the load kept repeating, "'Twun't ride!"

We had almost reached our destination when one front wheel of the wagon dropped into a rut. Slowly and inexorably the load started to go over. It had been so well built that it retained its shape almost to the ground and then dissolved into the vast, untidy heap that Fred had forecast.

Of Fred there was no sign. Someone called his name but there was no reply. We were really worried that he might have broken his neck and started frantically tearing away the pile of sheaves.

Right in the middle we came across a pair of hobnailed boots and, pulling away the last sheaves, discovered Fred sitting there glaring angrily.

"You alright, Fred?" asked the boss anxiously. Fred didn't bother to answer this stupid question, he merely said,

"Broke me bliddy pipe!" and then: "Told yer t'wouldn't bliddy well ride!"

But most days ended more happily. The black bulk of the rick loomed against the dusk and bats flittered jerkily around it, investigating the strange monster that had grown up while they slept during the heat of the day. I was young enough then to be able to hear the shrill squeaking of their echo-sounding, and the click and crunch of wing-cases as one of them found a luscious, night-flying beetle.

The boss drove the tractor back to the farm and I took Kitty out of the shafts and followed him. It had been a long day for the mare – bouts of furious exertion pulling full loads to the rick, between periods of fly-ridden inactivity when she stood haughtily surveying the humans as they scurried about her. Her pace quickened as we walked down the lane. She was looking forward to a small feed of rolled oats, a long drink of water from the stream; but most of all to being turned out with the cows on the Hangers.

Her routine was the same every night. Fred let her through the gate with a slap on the rump and, 'Goo' night, Gel. See yer in the mornin'.' She'd trot round in a big circle halfway up the hill, making the cows jump away from her in mock alarm as she established her superiority over mere bovines, then race down to her favourite level spot under the elm trees where she'd roll and kick her heels vigorously in the air for a minute or two. After that there was a brisk shake, from nose to tail, to get the dust of the day out of her coat – she'd stand looking at us for a moment as if to say, 'There, that's the way it's done!' – then down would go her head as she got busy on the task of re-fuelling ready for the next day's work.

And judging by the red glow in the western sky towards Guildford, another fine, busy day was promised us tomorrow.

10

OCTOBER

There had been one or two wet days during the last part of September, but even so, we had managed to get most of the harvest in before our gang of helpers departed – thankfully, I think – to their suburban houses and less strenuous 'business' occupations.

"My God!" said one of them surveying his hands, "they'll never be the same again." Broken fingernails, oil and grime ingrained, the tiny points of thistles lying hidden under the skin, and shiny, hardened callouses which had taken the places of the first few days' blisters due to constant friction on prong handles. We told him that a few days of pen-pushing would soon put matters right, and that his hands would soon be soft again. 'Til next year, that is, when we would get him back in training with some more 'real work'.

He looked doubtful at this, having obviously decided that there were more restful (if considerably more expensive) ways of spending one's summer holidays. But they had all worked wonderfully well and even Fred was sorry to see them go.

"Just beginnin' ter make summat of 'em," he said. "an' I s'pose next year'll be a diff'rent lot an' us'll 'ave ter start all over again."

They were good people and we really appreciated the help they gave us; and they enjoyed 'lending a hand on the land' when post-war austerity and strict currency regulations made package tours to Majorca still a thing of the future.

We only had one field of corn left at the beginning of October, but that was a bit of a problem. It was Browsy Field – as any field with a steep brow in it was always called – at the far end of the farm where Fulven's Lane joined the Holmbury St Mary road. Nine acres of late-sown barley full of charlock and thistles. We knew that it would be a real swine to cut with the old binder; the knife would clog in all the rubbish and the 'bull wheel', the big cleated wheel under the deck of the binder which drove all the moving parts, would lock solid and drag along ineffectually, tearing a furrow in the wet stubble.

"An' if us does get un cut," said Fred, "what 'appens then? That mucky old tackle'll never get dry enough ter stack proper. 'Tis getting' terrible late in the year ter expect 'arvest weather."

I mentioned the thistles, thinking of my hands when we were stooking, but that objection was brushed aside.

"Your 'ands don't matter, Boy, 'sides, yer ought ter be able ter take a few prickles by now. What do matter is that us won't never get that bliddy stuff dry enough ter put up a rick."

The problem was solved when the boss announced that there was a contractor with a combine in the area. He had been cutting wheat on the Downs near Dorking, and had agreed to send his combine to do our nine acres of barley the next day. As usual, the new idea was greeted with less than enthusiasm.

"Them combinder things mid be all right in America an' them places where yer got plenty o' room ter turn, but they'm gurt, awk'ard great things. Won't do fer ereabouts. An' what about all that loose straw what they do leave be'ind 'em? Yer can't do nothin' wi it."

The boss flared up.

"Hell's bells! What do you want me to do then? We can't cut it with a binder and now it's all wrong to get a combine in! If it was up to you lot the damn crop would rot in the field. No – I've made the arrangements and the combine's coming tomorrow." And he stomped off to the farmhouse to phone the corn merchant about hiring some sacks.

Maybe Fred had a vision of the future when all the skills he had practised for a lifetime would become obsolete when combines changed the face of the countryside – hedges grubbed out to cut out time lost turning at headlands; the complete absence of carefully-made corn ricks; and acres of straw being burned because combined straw is useless for thatching – but I was too excited at the prospect of seeing one of the wonderful new machines working to worry my head over the radical change that was taking place under my nose. Or perhaps Fred didn't realise it either and was just being his normal, cussedly old-fashioned self.

The combine created quite a stir as it came through the village to the farm next morning. It was followed by a respectful train of dejected-looking motorists who had been forced to stay behind the nine-feet-wide monster most of the way along the busy road from Dorking. The driver swung the machine onto the Marsh opposite the farm gate and directly the traffic had cleared, the boss, Fred and I went across the road.

Close to, the Massey-Harris 726, self-propelled bagger combine with its eight-foot cutterbar didn't look so enormous.

"I s'pose that thing be mostly a thrashin' machine, Boy," said Fred, "but 'e an't so big as the tackle they uses fer thrashin' ricks. Wonder what sort've sample 'e do turn out?"

"Look at the size of that knife, Fred." I said. "Cut a few acres a day with that!"

Fred grunted.

"Take some sharpenin', that would. Wear out yer bliddy file afore yer got ter the end of un."

The boss was deep in an anxious discussion with the combine driver who was sitting on one of the back wheels drinking tea out of a Thermos flask. They were both looking worried – the boss because he wanted to start cutting straight away, the driver because he didn't.

"It's the end of the season," he said, "I've been that busy I haven't had time to do any repairs."

"What's wrong with it?" asked the boss. "Will it work? Can you keep going long enough to do our nine acres?"

"Oh, it cuts all right," the driver replied, "it's just that I haven't got any brakes. It's not safe on a slope. I've got to change the brake shoes before I do any more cutting."

"Oh hell!" said the boss. "How long's that going to take?"

"Don't worry, Guv'nor. It won't take more'n half a day to get the shoes and fit them – a day at the most."

The boss looked terrified. Like every farmer, he knew that a day lost in autumn could mean that the barley would never get cut. If the weather changed it might rain until Christmas.

"Do you think you could possibly start right now?" he wheedled. "We're all ready for you – we've got the sacks and everything. It won't take long once you get started, and I'll make sure you don't get any stops...and I'll get my wife to fix you some tea and sandwiches...and cake at tea time... and I'll help you fix the brakes the moment you're finished...and...Please?

The driver looked thoughtfully at the grey sky, even more thoughtfully at his small dinner bag and then brightened up.

"Oh all right," he said, "I expect it'll be OK. So long as the field's fairly level."

"There be a gurt old brow in that there..." began Fred helpfully, but the boss shut him up.

"There's a little bit of a slant in one corner, but I don't expect you'll notice it on that thing. Follow Chris on the tractor and he'll show you where the field is."

While the driver started up his roaring machine the boss pulled me to one side and hissed in my ear,

"Take him up to the top end of the field and start him there, then he won't see how steep it is until he's started cutting. And for God's sake keep old Fred out of his way until he's got going."

We drove up to the Browsy Field, demolishing a narrow, binder-sized gateway or two on the way, and the combine started cutting. Fred went on the bagging platform and hooked sacks onto the grain spouts under the cleaner, switching the flow of barley from sack to sack and taking off the full ones. He tied them up quickly and expertly as if he'd been working on combines all his life.

"Have you done this before, Fred?" I asked.

"No, Boy, but 'tis just the same as a thrasher, an't it? 'Cept yer wouldn't be able ter 'andle them gurt, four-bushel bags up 'ere would yer?"

I'd wondered why we had got two-bushel sacks from the corn merchant – now I knew why. One hundredweight sacks were a manageable size to manhandle on a moving machine – two-hundredweight corn sacks would have been impossible.

I was supposed to be helping Fred, but only succeeded in getting in his way – he quickly had things organized, with empty sacks draped over the guard-rail near to his hand and a bundle of lengths of binder twine for tying up the bags – so I concentrated on watching the combine driver. He steered the machine one-handed, his left hand tugging at the knob on the steering wheel while his right hand was constantly moving, adjusting the height of the cutterbar, the set of the reel which gently brushed the standing corn back onto the knife, and most important of all, the lever which controlled the forward speed of the machine. Unlike a car or tractor, the speed of a combine cannot be varied with the throttle as the engine has to drive the threshing mechanism at a constant speed. So to maintain an even flow of corn into the thrashing drum the driver alters a variable pulley in the belt drive to the wheels, slowing down in thick corn and speeding up in the thin patches.

After we'd gone about fifty yards the driver stopped, allowed the corn in the machine to clear out of the back and then shut down the engine to idling speed. He climbed down and searched in the swath of thrashed straw behind the combine.

"Dropped summat, 'ave yer?' Broke the bliddy thing already?"

The driver grinned at Fred as he got back on the driving platform.

"No, old feller, I was just looking to make sure I wasn't blowing any corn out of the back. Sample all right, is it?" Fred dipped his hand into the sack and looked at the barley.

"Makin' a goodish job o' thrashin', she is," he said, "but there's a 'ell of a lot o' that old charlock seed in it."

"Can't do much about that. Charlock seed pods are the same size and weight as barley grains, so they're bound to come through in the sample. Still, you'll know what to do about it next year, won't you?"

"Yes I does." said Fred. "Cut the bliddy stuff wi' a binder an' thrash it proper from a rick."

"No," said the driver as he revved the combine up to working speed again, "don't grow so much rubbish in your corn then I'll be able to make a better job of it for you."

"Would a thrasher get those charlock seeds out of the barley, Fred?" I asked as we got going again.

"Might do, Boy," said Fred. "but not a lot better. I were just 'avin' a go at 'im. No – 'e's doin' a good job considerin' 'ow dirty this corn be. 'E knows what 'e's up to, that lad does."

Coming from Fred that was very high praise indeed. Although the 'lad' was only a few years older than I was, his competence in handling the combine had already earned him the status of craftsman in Fred's estimation. And while Fred's skills of rick-building, thatching, hedge-laying and horsemanship would probably be a closed book to the combine driver – Fred would be equally at a loss if asked to operate a complicated machine powered by forty mechanical horses. Craftsmanship in farming doesn't change, but the methods and techniques do.

When he got to the top of the brow in the field, the driver saw how steep it was. He shouted back to the bagging platform.

"Little bit of a slant? That's a ruddy mountain!"

"I tried ter tell yer," Fred replied, "but yer wouldn't listen. Cut across un, like us would wi' a binder."

"No good," shouted the driver. "it's too much of a side slope. She'd tip over. I'll have to cut it up and down."

He went on, easing the big machine down the slope very slowly, the engine and transmission doing all the braking. All went well until there was only about half an acre left to cut, right on the steepest part of the brow. Then it happened. One of the two driving chains snapped under the unaccustomed load and the combine shot off down the hill like an express train. I jumped off but Fred stayed on the bagging platform, clinging to the rail. I sat on the stubble and watched the machine as it streaked towards the roadside fence at about thirty miles an hour. The driver managed to pull the combine round just before he reached the fence, and it cornered on two wheels, coming round in a big circle and gradually losing speed on the level ground.

The boss had been watching from the brow and we ran down to see if anyone was hurt. The driver, white-faced, climbed shakily down and sat on the step. Fred stayed on the bagging platform, quite unperturbed, grinning hugely as he stuffed black shag into his pipe.

"See the ride I 'ad then boss, Boss, did yer?" he called cheerily. "She don't 'alf go, don't 'er!"

The next day the driver repaired the broken driving

chain and fitted new brake shoes to the combine. Then he drove it home. He didn't finish the half acre of barley on the steep brow and eventually we ploughed it in. But as Fred said,

"'Twere mostly all thistles an' charlock, anyways, Boy. An' the birds 'as 'ad most on it."

I was a bit disappointed that there was no rip-roaring harvest supper held at the farm when 'All Was Safely Gathered In'. I had read about wild junketings in decorated tithe barns; long trestle tables groaning with home-cured hams and joints of English beef; red-faced, Sunday-suited yeomen waving foaming mugs of strong ale as they roared out traditional country songs; and similar excesses in celebration of the battle of the harvest having been fought and won for another year. But there was nothing like that. The boss took us to the pub one evening and bought pints all round. We dutifully discussed the harvest, laughing over some of the funny things that happened – "Remember that big load that came off, Fred? You must've been doing about forty miles an hour when you hit the ground!" – and then the topic of conversation turned to the recent christening of Princess Anne, the new daughter of Princess Elizabeth and Prince Philip – King George's first granddaughter. Fred and I soon tired of that and went and played Shove-Ha'penny. When we looked round the boss and Ernie had gone home. Not much of a celebration, but in those days of austere Socialism, better than nothing.

The farm soon went back to its normal routine. Fred was busy with another of his 'expert' jobs – thatching the hayricks that weren't going to be started until the late winter and putting a thin layer of thatch on the corn ricks that would be thrashed out fairly soon, depending on when the thrashing contractor from Cranleigh got round to us.

I found myself helping Ernie more often with the cows. I was a competent hand-milker by then, thanks to Ernie's strict training in the cowstall, and I soon picked up the routine of machine milking on the outdoor bail. It wasn't long before I was trusted to cope with the herd on Ernie's Saturday afternoons off. I was immensely pleased with this bit of promotion and thought I was making meteoric progress up the farming ladder – why, less than a year before I hardly knew which end of a cow you got milk out of, and here I was, now, milking a whole herd single-handed. It didn't occur to me that I was left to do the milking because the boss wouldn't have missed his Saturday afternoon rugger match for the world. I wasn't supremely capable – I could just about manage – and Fred had been instructed to keep an eye on me.

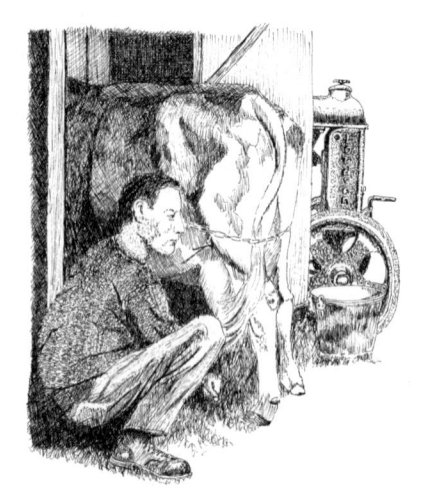

Fred had the farmworker's natural mechanical aptitude and could handle complicated machines like a binder or thrashing machine quite competently. He had been using milking machines, he told me, right back in 1922 when he was an under-cowman on a farm near Guildford. But he couldn't do the relief milking as he had a mental block when it came to internal combustion engines. He refused to have anything to do with them, apart from complaining loudly about their fallibility compared to horses when they broke down. He couldn't drive the tractor used to carry churns and feedingstuffs to the outdoor bail, and the $1^{1}/_{2}$ hp petrol engine that drove the milking machines had him stumped completely.

It had me stumped, too, one wet Saturday afternoon, and if it hadn't been for old Ted, one of the 'characters' of the village, I wouldn't have got the milking done at all.

Ted was a pensioner and spent all his waking hours, when it was fine, sitting on the low wall outside his cottage next to the post office. He rolled cigarettes with a practised hand, spat accurately into the gutter and chatted amiably with passers-by. His place on the wall was an essential part of the village economy because everyone knew that he would always be on duty there, ready to help them out in times of trouble. For Ted was the village mechanic.

He came out of the army at the end of the 1914–1918 war with a smattering knowledge of engineering, great enthusiasm for life and a natural-born genius with anything mechanical. Take a clock, lawn mower, bicycle, car, tractor, or any sort of farm or garden machine along to Ted and his eyes would light up with the thrill of having a problem to solve. He would run his hands lovingly over the faulty mechanism or rough-running engine and tell you exactly what was wrong with it. I was always impressed with Ted's method of testing the ignition on Fordson tractors. Fordsons had very powerful impulse magnetos which, if you inadvertently touched one of the leads, gave you a jolt that made your teeth rattle. Ted would take hold of each plug terminal in turn while the engine was running to find out which cylinder was misfiring. "Bit weak on number three," he'd say. "I'll stop 'ar and clean the mag points." Then he'd stop the engine by shorting-out all four plugs with both hands.

And it was no good trying to take anything away from Ted once he'd diagnosed a fault, he'd only return it to you when it was repaired and running again. The modern practice of fitting 'Factory Reconditioned Units' to replace worn or broken components revolted him. "There aint nothin' made what can't be mended," he said.

The proprietor of the local garage hated Ted with a deep and bitter loathing. Not only did Ted do his repairs very cheaply – he would never accept more than five shillings or an ounce of tobacco in payment – but he couldn't help interfering with what they were doing at the garage. Ted had been the original owner of the garage which he started up after his army service. He rented an old stable and turned it into a workshop, set to and made himself a set of tools because he couldn't afford to buy them, and was soon flourishing. Model 'T' Fords and Austin 'Sevens' needed a skilled and understanding hand to keep them running, and Ted's hand-wound petrol pump supplied them with fuel. His pride and joy was an ancient Fiat lorry belonging to Mr King, the local builder. It had come to this country by some devious route from the battlefields of France, and soon after its arrival the engine blew up. Obtaining spares for a foreign lorry in the early Twenties was – to say the least – a little difficult, so Ted got busy. He rivetted the shattered oil sump, straightened and repaired the mangled engine parts and put it all together again. It ran beautifully and was still going strong when I knew it thirty years later.

Just after the '39–'45 war Ted had a bad accident. He was repairing a generating set at Paddington Farm and his trousers got caught in the unguarded driving belt. He lost a leg but, typically, he didn't lose consciousness while they were waiting for the ambulance but demanded a cigarette. When they gave him one he said, "There you are, you see! I keep tellin' you what happens when you're bloody careless, don't I? Good job it were me an' not one o' you young fellers."

It wasn't long before Ted was up and about again, proudly showing everyone the mechanical marvels of his new, artificial leg. But he had to sell up the garage business and made the new owner's life a misery by criticising everything he did.

I went to the milking bail that Saturday afternoon, filled the feed hoppers, opened up the canopy, let the first four cows in and went to the cupboard which housed the engine. I swung the handle until I was dizzy, but there wasn't the slightest whisper of a cough from the exhaust – the engine was absolutely lifeless. I didn't know what to do. Then I remembered old Ted. I let the cows out again, shut the bail down and drove the tractor up the road to the post office.

When I got to Ted's back door, he was struggling into his raincoat.

"'And me my 'at, Missus. Where's me stick? I shan't be long – boy's got trouble wi' 'is milkin' engine."

"How did you know?" I asked, wondering if the old man was a mind-reader. I hadn't said a word.

"'Cause you got churns on your trailer," he replied, "An' you wouldn't call me for nothin' else on a Sat'day. Got petrol in it, 'ave you? Got a spark? Shouldn't be the magneto – I done that last week. It'll be a valve stuck I reckon."

I carried his toolbox out and helped Ted up onto the trailer, then we drove back to the field. When we got to the bail he eased himself stiffly down and opened the engine cupboard. Rain poured off the roof onto his back but he didn't seem to notice. He crooned over the engine.

"What's wrong wi' you then? Don't want to start then, do yer? Let's give you a turn over, shall we, an' see what's up. Ar, that's it. You 'aven't got no compression, 'ave yer? You've got a valve stuck, 'aven't yer?" Without turning round Ted reached an open hand towards me. "Half-inch ring spanner, Boy. In the top o' me toolbox."

I slapped the spanners into his hand as he demanded them, like the operating theatre assistant attending the great surgeon. And within a few minutes the patient's innards had been exposed, a bit of skilful open-heart surgery had been performed on the offending exhaust valve, all was closed up again and the recovery was complete. One swing of the starting handle and the engine puttered into life.

Ted refused to let me take him home on the tractor, saying that I was late enough milking as it was, but limped off through the mud, swinging his gammy leg high at each step. "That's alright, Boy, I can manage. You bring my toolbox home after milkin', mind, and tell your boss 'as 'e owes me a ounce o' baccy." A cheery wave of his stick through the rain and he was gone. Next day I told Fred all about the engine breaking down. "But I managed to get it going in the end" I lied. Fred knew.

"I just 'appened to be in the 'llotment a-watchin' yer start milkin'," he said, "when I see yer a-cussin' an' a-kickin' at that engine what wouldn't go. I know'd yer 'ad trouble an' I were jus' goin' ter come over an' give yer a 'and, like. Then I see yer go up along fer old Ted an' I knowed yer'd be alright. Proper marvel, Ted is, wi' engines an' that."

"Yes, he is." I said. "The boss ought to have got him to fix the brakes on that combine, then it wouldn't have run away in the Browsy Field."

"Ar, that's right." Fred chuckled reminiscently. "But then I wouldn't 'ave 'ad my excitin' ride down that old brow, would I? Tell yer what, Boy," he said, "now I knows what 'tis like a-flyin' in one o' they airyplanes!"

11

NOVEMBER

Most of the villages around Abinger had firework displays and bonfires at the beginning of November – there was always a big one at Shere to which people use to flock from miles around – but the most exciting one was the one we had at the farm. It wasn't our celebration of the anniversary of Guy Fawkes's attempt to dismember Parliament, but quite accidental and unplanned. The cause of it was the boss's car, a pre-war Ford 'Popular'.

Like most farm cars, before the days when farmers drove to market in Mercedes and Jaguars, the boss's Ford wasn't an everyday 'social, domestic and private' vehicle, it was a 'farmer's goods and private' – a farm implement. And it looked it. The body – originally shiny black – was the colour of dried mud because that's what the outside layer was –

dried mud. The inside being much the same. The back seat had long since disintegrated under the impact of calves, piglets, the boss's big black dog, Tex, rolls of barbed wire, drums of oil, and sundry sharp pieces of ironmongery such as plough shares and mowing-machine knives. The latter, being too long to go across the seat, were carried with the heel – the blunt end – sticking out the left-hand window while the razor-sharp sections at the other end wrought havoc with the seats and what was left of the interior trim.

But on the whole, once it had got going, the car went well and could achieve speeds in excess of thirty miles an hour on level roads. The main difficulty was persuading it to start from cold – it had the old Ford side-valve engine, the design of which hadn't radically altered since the Model 'T' – and the starting procedure was a complicated and hazardous operation – the direct cause of our fire.

The battery of the car had expired some years before, and the only way of bringing the engine to life was by energetic use of the starting handle. And to get it to fire, the engine had to be 'primed' by pouring neat petrol into the carburettor air intake. More often than not this resulted in the engine becoming hopelessly over-choked, in which case the plugs had to be taken out and dried on the gas stove in the farmhouse kitchen, but sometimes it worked. And on the memorable day of the fire it worked rather too well.

The boss told us afterwards that he had gone to start the car, which was parked in the tractor shed, in quite the normal way. He opened the bonnet, took the cap off a two-gallon can of petrol and poured some into the carburettor. Then he went round to the front of the car, rolled up his sleeves and prepared for the long job of turning the handle rapidly and continuously until the engine fired. And fire it did – or rather, backfired.

The first time he turned the handle there was a loud bang, the handle spun out of his hand and flew backwards, and a sheet of flame erupted from the carburettor, showering the engine and the wall of the shed with burning petrol. The open can of petrol was standing alongside the car and that added its contents to the inferno. It all blazed up and in the close confines of the tractor shed the roar was terrifying.

The boss's first thought was not the car – that was insured and I think he would have been rather pleased to have seen it reduced to a charred 'write-off' – but the hundred gallons or so of tractor paraffin standing in barrels just behind him in the shed. Also the fact that the tractor shed was made of wood, leaning-to against the wooden stable with its 'tallet', or hay loft above it. So he strained every nerve to push the burning car out into the open yard, shouting desperately for Ernie, the cowman, as he did so. Ernie came running from the dairy and the boss's wife, hearing the commotion, looked down from the farmhouse, saw the flames and smoke and phoned for the fire brigade.

Ernie had been washing down the passageway from the dairy and the hosepipe was still on the tap. He turned it on and ran to the full extent of the hose, just managing to reach the roof of the tractor shed with the thin stream of water.

The boss dipped a bucketful of boiling pig swill out of Fred's copper by the pig yard – it was the nearest available liquid – and slopped the greasy mess over the engine of the car. Then he went back for another bucketful for the near-side front tyre which was blazing and spluttering. Fred's pig swill must have been pretty potent stuff because the flames went out immediately, leaving the front of the car steaming gently and giving off a thick stench of burnt oil, burning rubber, fried potatoes and stale cabbage water.

Ernie had finished soaking the roof of the shed and turned his hose on the flames inside. They had started to die down a little as the petrol that had started the fire burnt away. And the paraffin, leaking from a barrel that had split in the heat, had run across the beaten-earth floor of the shed and down a drain just outside the door. The drain was full of water and the flames went out. Ernie's hose quenched what was left of the fire and, as quickly as it had begun, the fire was extinguished. We discovered afterwards that apart from the damage to the car, the losses were two gallons of petrol and about fifty gallons of paraffin. The walls and the roof of the shed were blackened and steaming, but thanks to Ernie's prompt action with the hosepipe, the wood hadn't caught and the flames hadn't spread through the wall planking to the hay and straw in the stable.

The two men were congratulating themselves on their escape – the boss nursing his arm where it had been scorched when he pushed the burning car out – when the clamour of a fire bell cleared the way for a fire engine round the sharp corner in the village. The big, red Dennis engine roared into the farm gateway and almost before it skidded to a stop, the firemen had unrolled hoses and started the pumps.

"Is it a fuel fire?" shouted the fire chief. The boss nodded.

"Yes – but we've got it..." The fireman didn't give him time to finish.

"Foam, Lads!" Then to the boss and Ernie, "Out o' the way please, Sir. And turn that hose off – we don't want water."

Although the fire was out, the firemen made a thorough job of covering everything in thick foam. Then, when they had finished, they went all over it again – 'damping down', they called it. They were so keen they even smothered the dung-heap outside the stable which was minding its own business and steaming organically in the way that dung-heaps always do.

Eventually the fire chief called a halt.

"Right, Lads, that'll do. That's got 'er." Then he walked across to where the boss and Ernie stood by the car.

"Good job you called us, Sir. That could have been nasty."

"Thank you very much," said the boss, "but actually it was Ernie with the hose that..."

"Ah yes," the fireman interrupted him again. "You want to have a word with your blokes about using water on a liquid fuel fire. Worst thing you can do! Spreads it in no time, it does. It makes the fuel vaporize and explode."

The firemen had rolled up the hoses, shut off the pumps and closed the panels on the fire engine. Now they gathered round their chief, taking their helmets off and wiping their brows with exaggerated gestures. They nodded in agreement.

The boss looked at them helplessly.

"Oh well, I suppose you know more about fighting fires than we do. And talking about liquids – how about some beer?"

They all nodded again happily and clumped up to the farmhouse in their heavy boots. It was obvious that they had forgiven the boss for his ignorance of firefighting methods – he did, at least, know the correct way of damping-down afterwards.

Another exciting event in the month of November, and planned this time, was the annual 'thrashing' as the operation of threshing corn ricks was always called. In Hampshire and Wiltshire it was still called 'sheening', a contraction of 'machining' and, one presumes, originally a derogatory term to describe the invention which did away with the farm labourer's main source of income during the hard months of winter – thrashing out corn with flails on the barn floor.

Thrashing machines are large and expensive and used for only a few days in the year, so it was only the big farms and estates who could afford to have their own set of tackle. The relatively small farms of Surrey used to rely on the services of contractors who owned the machines, their attendant balers and straw trussers (straw for thatching has to be bundled up in loose trusses rather than being broken and compressed in a baler), and the large and powerful tractors needed to move the thrasher from farm to farm and power the whole outfit when on site. Our local contractor was Weller & Son, from Cranleigh.

One disadvantage of not having your own thrasher was that it was always more expensive to handle your crop. Very often, in a good season, the owner of a thrashing set could thrash his own corn straight from the field – carting the sheaves from the stook to the thrasher and thrashing it out there and then – saving the trouble and expense of building and thatching corn ricks and getting the crop away and sold sooner. This meant that the following autumn's sowing could be financed by money in the bank rather than expensive borrowing for seed and fertilizers while one's own capital was tied up in corn ricks waiting to be thrashed.

Another reason why the smaller farmer could not run his own thrashing tackle was that the job needed eight or nine experienced men – almost double the usual labour force. So neighbouring farmers helped each other out on a co-operative basis, each one sending two or three men to help with the other's thrashing and having the benefit of a full gang when the time came to do his own. The arrangement worked well and we enjoyed working on other farms where the difference in routine and methods made a welcome break. Also it enabled us to indulge in a farmworker's favourite pastime – criticising and poking fun at the neighbours.

The first farm the thrasher called at in our group was in Gomshall owned by a retired army colonel. Everything ran like clockwork and the whole farm was always in apple-pie order. Rumour had it that the colonel had been retired early from the Royal Military Police because he was over-zealous in the matter of discipline when commandant of an army detention barracks, but the tales were probably grossly exaggerated. Working farmers and their men just naturally despised (and envied) those who could afford to buy a farm, put up brand-new buildings, buy new and wonderful labour-saving machinery, and start a first-class herd of pedigree Jerseys by using a cheque book rather than years of hard work and selective breeding. The trouble was that a lot of these 'gentlemen farmers' – and I don't think the colonel was one of them – designed their farming enterprises to run at a loss, so that the losses could be set against their income from business and other interests quite outside farming for tax purposes. Before the tax laws were changed, putting an end to this sort of tax avoidance, it was deeply resented by 'real' farmers who had to struggle to meet the tax bill from their farm incomes. Nobody bothered to mention the vast amount of wealth that was being poured into farming as a whole from the captains of industry who liked playing at being the squire at weekends, but it must have been considerable. Farm machinery, for example, would have been much more expensive had not the wealthy men been buying, and the makers couldn't have spent so much on research and development of new machines.

But one criticism of the influx of businessmen into farming – particularly in the Home Counties – that they were buying up farms which had formerly been rented, was quite justified. Between 1950 and 1960, three million acres of land in England and Wales changed from being rented to being 'owner-occupied'. This meant that at an average of 150 acres, there were 20,000 fewer holdings available for farmers' sons and aspiring farmworkers who wished to start farming but lacked the large amount of capital necessary to buy a farm of their own.

So criticisms of the 'gentleman farmers' abounded and any mistakes they made were crowed over and judged to be a result of their incompetence and ignorance of real farming – as opposed to text-book farming; and their successes usually went unremarked except where it might be said grudgingly that something or other 'wasn't too bad'. In these cases the credit for any successes was given to the gentleman's long-suffering staff who, while they were sneered at just as much as their employer, were generally considered to be working 'in spite of' the owner, and not 'for' him.

While all this was going on our military neighbour was a good farmer and treated his men well. They seemed contented and stayed with him, but they had to toe the line and keep everything in order. If they didn't there was trouble.

The colonel rang our boss one morning at seven o'clock.

"Spot of bother, Old Boy," he said. "Need your assistance."

The boss glanced at the clock and jumped to the obvious conclusion.

"Are my heifers in your kale again? I'll come right down and fetch them."

"Not your cattle – mine." replied the clipped voice of the colonel. "Trouble with the Jerseys. No-one to milk 'em. Do 'em myself but I don't know how. Bit adrift when it comes to teats and things, y'know."

"Is your cowman ill then?"

"Not ill – useless." the colonel replied. "Damn feller kept comin' in late. Warned him again yesterday. Twenty minutes. Unpunctuality. Damned bad, y'know. Won't do. Couldn't put up with it. Had enough."

"What happened this morning?" asked the boss. Didn't he come at all?"

"Yes he did. Chap was twenty minutes late again. Not sorry in the slightest. No excuse. Damned insubordinate in fact. Gave him his cards. Sent him orf. Minute's notice."

"I'll come over and give you a hand then," said the boss," but I think you were a bit hasty. You should have waited 'til after milking and then sacked him."

"Damned grateful," said the colonel. "very civil of you. Had to do it there and then, though. Decision was made. Not fair on the man, y'know, to return him to duty and then let him go afterwards. Principle of the thing. Fair's fair. Always be straight with 'em. Works with the other chaps. Worked in the army, dammit!"

The boss told us all about it after he'd been down and helped the colonel's tractor driver milk the Jerseys. "The colonel was right to get rid of that cowman of his. Bone idle! The yard and milking parlour were in a hell of a state. I wonder the old boy put up with it as long as he did. And, of course, he didn't have any idea about milking. He's never had to do anything himself – all he's ever had to say was 'Carry on Sarn't' and things were done for him". He paused for a moment and then said thoughtfully,

"It was strange being down there this morning. Like being back in the army again."

"Make yer jump about a bit did 'e, Boss?" said Fred. "'Twouldn't do fer I, workin' fer no colonel. I'll wager 'tis all 'Squad Shun!' an' 'Get yer 'air cut!'"

We walked across the fields to join the colonel's thrashing gang and Fred talked about the time he was in the army during the 1914–1918 war. He'd never mentioned it before and I hadn't raised the subject, thinking he might have memories of the trenches in France which he didn't like to recall. The boss was the same. He'd been right through the 1939–1945 war and didn't talk about the horrors of it – only the amusing bits. For example, the only story he told us of the gory fighting up through Italy was of waking up one morning and finding himself sitting at the wheel of a jeep parked outside the front entrance of a palatial villa. "There was as enormous flight of steps leading to the gardens below," he said. "Hundreds of them – all marble. We'd had a bit of a piss-up the night before and the only way I could have got that bloody jeep up there was to have driven it up all those steps in four-wheel drive. To this day I can't remember doing it!"

But in Fred's case, he didn't talk about life as a soldier because it was the only time he wasn't farming – the only interesting thing to be doing. And he brightened up when I asked, tentatively, whether he'd ever been on active service.

"I bliddy well were an' all, Boy," he said. "I were two year drivin' army 'osses on timber carts down Sussex. That were real active, that were." Then he said,

"Come on, Boy, best step out. Us better not be late on p'rade at colonel's thrashin', or 'e'll send us 'ome double quick an' that won't get no work done."

When we got to the colonel's ricks I was amazed to see that everyone was smoking. I asked one of the colonel's men why it was allowed, as it had been drummed into me very early on that you never smoked near hay or straw.

"The guv'nor 'as to 'ave 'is pipe goin' all the time," he said, "can't do wi'out it. An' 'e says as 'ow 'e aint goin' ter stop us doin' summat 'e does 'imself. E's a fair man, 'e is."

He pointed upwards and there was the stern disciplinarian standing on top of a corn rick puffing happily at his pipe – breaking one of the strictest rules in farming. Fred wasn't impressed.

"I likes me bit o' bacca," he said "but I an't goin' ter smoke round no ricks – 'taint right. An' you an't goin' to be neither, Boy. Just 'cause we be away from 'ome, like, you an't goin' ter start no bad 'abits. 'Tis wrong."

When the thrasher arrived at our ricks it was like the circus coming to town. At the head of the procession was the big, dark-green, single-cylinder Marshall tractor, its huge flywheel spinning away alongside the engine cowling and black clouds of diesel smoke pumping out of the broad exhaust stack in time with the staccato bark of the engine. It hauled the vast bulk of the thrashing machine, and hitched behind that was the long, four-wheeled Jones wire-tying stationary baler, and behind *that*, a smaller, two-wheeled string-tying trusser for bundling thatching straw – and as if that wasn't enough, finally at the end of a towrope, steered by the thrashing contractor's weasel-faced mate, the contractor's small Ford car. He used this car to go home to Cranleigh each night and towed it behind all his tackle when moving between jobs to save his rationed petrol.

It took about an hour to set up all the thrashing tackle between the two ricks. The job was made more difficult because the ground was wet and slippery and there was a bit of a slope where the thrasher had to stand – the five-ton machine slid down the hill on its broad rubber tyres the first time the contractor tried to tow it into position. But he was used to bad conditions and the Marshall was fitted with a large and powerful winch. He paid out the wire cable, hitched it to the drawbar of the thrasher and then, from twenty yards out in the field, pulled the wooden monster at an angle uphill of where he wanted it to go. It was a joy to watch the thrasher as it slid sideways again, but this time into exactly the right place. The contractor did the same with the baler but this time he had to be even more precise. The baler was driven by a belt from the thrasher and the two pulleys had to be exactly in line with each other or the belt would fly off. And the distance between the two machines had to be right, too. Too close and the belt would be slack and wouldn't drive properly, and too far away and it wouldn't reach.

I remember one occasion when we were going to thrash a rick of beans. The ground was very wet and, try as he might, Mr Weller couldn't get the baler any nearer than three yards away from the thrasher. He hadn't got a long enough belt to take the drive from his thrasher, but he solved that problem by dashing off to his small car. He returned an hour later driving a Fordson Major which had a belt pulley fitted to it. He drove the tractor round to the other side of the baler, slipped the driving belt on and we were all ready to go, the Fordson Major driving the baler separately while the Marshall drove the thrasher as usual.

I had been eyeing the three-yard gap between the two machines and a thought occurred to me.

"The straw isn't going to fall into the baler, is it?" I said.

"My God, you're quick," said the boss. "we'd never have noticed that if you hadn't pointed it out to us."

Fred leant on his prong and spat on the ground under the straw-walk of the thrasher.

"Round about there, Boy, that's where it'll drop."

"That means that somebody's got to stand there all the time and pitch the straw up into the baler."

They all gave big sighs of relief and Mr Weller grinned as he wiped his hands on an oily rag.

"I was wondering who you were going to put on that dirty old job, Bob." he said to the boss. "Now we've got a volunteer." The boss patted me on the shoulder.

"Keen as mustard, Chris is. Good thing too – he'll need to be on that job. Come on, let's get started."

Bean straw is as heavy as lead and covered in filthy black dust. I staggered home that night dog-tired. Instead of grumbling about the state of my clothes my mother burst out laughing when she saw me.

"I haven't seen anyone as black as you are since I was in India," she said.

When the baler and the thresher had been lined up by those first two ricks a lot of time was sent in levelling-up the thrasher. Holes were dug for the upright wheels and the downhill side was jacked onto blocks of wood, until the spirit levels built into the side and end frames of the chassis showed that the machine was perfectly level in both directions.

"They 'as ter be like that so's they thrashes proper." said Fred. "Gawd knows 'ow them combinder things do manage when they be runnin' so bliddy unlevel most o' the time."

When the job was done to the contractor's satisfaction, the Marshall was lined-up with the thrasher and its wheels securely chocked so that it wouldn't move forward with the pull of the belt. You never apply the brakes of a tractor that's on heavy belt work as the rocking motion tears the brake linings to pieces. And when all was ready we started thrashing.

A squeal from the clutch, a burst of smoke from the exhaust, a sharp cracking roar as the big, two-stroke diesel opened up to full power, and the thrasher's drum growled up to a steady, humming drone. The contractor went round carefully, inspecting the many driving belts and moving parts, checking that the baler was running properly, then nodded to his weasel-faced mate who was perched on top of the drum.

Fred had already stripped the thatch from the rick and stood on the peak of the roof with his short prong. He pitched sheaves to me where I stood alongside the top of the thrasher, pitching them on to Weasel-Face who was the 'bond-cutter'. He caught each sheaf as it came to him, cut the string with a sharp, hooked knife, and with a practised movement, spread the corn into the thrashing drum in an even flow.

We had one stop during the morning for a tea break, while the contractor went round the thrasher and baler with an oil can, feeling bearings to see if they were running hot, and checking the joints in all the driving belts. We swigged at cold tea – without milk and sugar the most refreshing drink there is on a dusty job – but Mr Weller had his steaming hot. A few minutes before he stopped the engine of the Marshall he unscrewed the plate on top of the tractor's radiator and lodged his bottle of tea in there to heat, steam whisping round it and coiling away in the draught of the cooling fan.

Fred told me that in the old days beer and cider was provided by farmers during haymaking, harvest and thrashing.

"That must have cost the farmers quite a bit." I said.

"No it never, Boy, leastways – they got their money's worth out o' it. It were reckoned ter be part o' yer wages in them days. Never got no proper overtime an' that like us does now – got the drink 'stead o' it."

I asked him whether many men got drunk in the fields and he shook his head.

"If yer got fuddled in the 'ead an' couldn't do no work, yer got sent 'ome an' never got no wages. But I remember one bloke that was doin' 'is job, bond-cuttin'" – he pointed at Weasel-Face who was having a smoke away from the thrasher –" 'e 'ad a drop too much, lost 'is balance an' fell in the bliddy thrashin' drum."

I shuddered. "How horrible!"

"'Twere 'is own fault." said Fred who always regarded the farming operation as the most important thing. "Trouble was us lost two days thrashin' while they cleaned the bits of un out the bliddy machine."

We soon started work again and as the level of the corn rick had gone down below the feeding platform of the thrasher, I found it more difficult to pitch the sheaves to

Weasel-Face's hand. He nagged at me constantly over the hum of the thrasher.

"Not that way, Boy!" – "I'm over 'ere, I am, not 'tother end o' the soddin' drum!" – "Fer Gawd's sakes pitch 'em a bit sensible like!" – "Blimey, Boy, you're bloody useless, you are!" and so on. Every sheaf I pitched up brought a sneer until I got flustered and desperate, flinging the sheaves up wildly. Eventually Fred walked across the rick and took hold of my arm.

"Stiddy up, Boy, yer won't get nowhere like that. Pitch 'em up quiet an' easy like I showed yer. Don't take no notice o' 'im – 'e's like that wi' everybody."

When dinner time came Fred and I sat in the shadow of the thrasher and ate our sandwiches. Weasel-Face came and sat beside Fred. I pretended to ignore him, feeling hurt and irritable as only a sixteen-year-old can. For a while the two old men chatted about this and that, the state of the harvest that year, the weather, and how each farm's crops had yielded. Fred was always interested in the goings-on on other farms and Weasel-Face, who travelled over a wide area with the thrasher, had plenty of stories to tell him. But I gathered that they liked coming to our farm as the corn was always in good condition, the work was organized efficiently and – unlike some – our boss paid his thrashing bill promptly. Then Weasel-Face leaned across Fred and spoke to me.

"How you getting' on, Boy? You don't say much."

"Quiet's a mouse, an't 'e?" said Fred. "Gen'ally talks all the time, 'e do too. Reckon you've 'ffended 'im wi' yer bit o' schoolin' of 'im up on the rick."

"Ar – that's it, is it?" said Weasel-Face. "Well, there were a reason fer that, Boy."

"What reason?" I said crossly.

"'Cause you was pitchin' all wrong when yer started. I kep' on at yer 'til yer got it right. An' ter tell yer the truth I likes ter be a bit evil wi' you youngsters – pushes yer, I does, ter see if yer got the guts enough ter put up wi' it. You'll do, you will, now that I've a-learned yer." He pushed his sandwich tin towards me. "'Ave a bit o' cake? Me name's Tom."

I'd been accepted and passed another exam.

"Thanks, Tom," I said. I took the biggest slice I could see and bit into it. It wasn't nice but I didn't say anything. Fred would have thought it extremely bad manners to have criticised a peace offering.

12

December

The big Marshall tractor chugged away down the lane towing its long train of thrashing tackle. Where the corn stacks had been there were neat stacks of rectangular wire-tied bales of straw; shaggy stacks of trussed straw – wheat for thatching next year's stacks, and soft oat straw for feeding – and vast, untidy heaps of cavings, dust and weed seeds. And in the farmyard, every inch of spare space was filled with the fat, brown bodies of four-bushel corn sacks, each one weighing 2¼ hundredweight if filled with wheat, 2 hundredweight of barley or only 1½ hundredweight of the much lighter oats. They had to stay in safe, dry storage until the corn merchant's lorry came to collect them, and they were lined up very carefully. No sack was ever allowed to rest on the cobbled floors of the buildings or moisture would have come up from the earth underneath – the builders of the old farmyards didn't put in damp-proof layers – and all the sacks rested on slats of wood or poles so that there was an air space under them. Each line of sacks was six inches away from the wall or the next line, so that the farm cats could patrol up and down and prevent the suddenly bountiful supply of food being colonised by rats and mice.

"There's a whole year's work in them sacks, Boy." said Fred. "You got ter store un real well. Each sack got near enough three quid's worth in 'im. Bliddy near a week's wages!"

I asked Fred whether the grain would keep all right with the sacks tightly sewn up with binder twine – rolled up along the top with a big ear sticking up at each end – and he had a sneer at combines by way of an answer.

"That'll store bliddy lovely, Boy. Corn what's come off've a thrasher be as dry as a bone. Not like that wet old tackle you gets off these combinders. You go ter leave the sacks open for they – an' keep a-turnin' 'em – or that'll be as mouldy as last week's cheese in no time."

The boss spent quite a lot of time inspecting the sacks of corn, counting them up and scribbling calculations on the backs of old envelopes.

"Cor, booger, 'e's in there again countin' them bags," said Ernie. "I swear that's the third time today. I reckon as 'ow 'e 'opes there'll be a few more each time 'e counts the bliddy things."

"He's got to sell quite quite a lot to pay the corn merchant's bill, hasn't he, Ernie?" I said.

"Long as 'e don't sell it all," Ernie replied. "I shall want a good bit ter make a decent old mix for them cows o' mine this coming winter. Course – 'e can sell *some* on it," he said generously, "but I got ter make sure 'e buys me a bit o' cake instead. You can't get milk out o' one end of a cow 'less you puts summat in t'other end." And he stumped off to do his work leaving me thinking what a complex business farming was – selling a bit of this to buy a bit of that – and always having to be a year ahead with your planning.

Fred and I spent a day or two at the places where the ricks had been, clearing up the old rick staddles and setting fire to the heaps of cavings. They smouldered slowly, giving off a rich bonfire smell for many weeks, sometimes until well after Christmas. But when they had finally dwindled to small heaps of white ash, all the weed seeds brought in with the corn had been destroyed. Except, of course, where the combine had cut the crop. Fred was quick to point out to me, exactly in the lines of the swathes left by the machine, weeds were growing thick and green.

"See that, Boy? Use one o' they combinders an' very soon yer farm'll be over-run wi' rubbish. Never gets that wi' a binder, yer don't – they lets yer cart yer weeds off the field 'stead o' plantin' 'em again for yer."

He was right there. Some weeds, like poppies, had been tolerated in the fields because they never reached epidemic proportions. They added immensely to the attractiveness of the harvest scene, and their thin scattering was due to relatively few seeds having been left in the stubbles the previous autumn when the sheaves were carted off. But when the combine had passed by, blowing *all* the weed seeds back on to the stubble they all germinated, or worse still, lay dormant waiting to 'increase an hundredfold' in the next year's crop, instead of being burnt in the stackyard. To combat this massive invasion of weeds, farmers had to resort to chemical 'selective' weedkillers which, far from being selective, kill everything except the crop including the wild flowers in the hedges. All due to the coming of the combine.

Another change that was taking place on the farm was the cutting out and clearing of the vast rews that grew along the boundaries and in between the fields. The word 'rew' is a dialect word which probably came down from the Old English word meaning 'rough' – and rough is a good way of describing the hedges before the boss took the farm. As Ernie and Fred said, "Them old rews han't felt iron fer nigh on twenty year." They spread out, a tangle of hazel, elm tillers, elder and thorn, a good twenty feet from the original line of the hedge, an effective barrier against cultivation, but useless as a barrier to livestock. Unlike a properly-tended hedge they had grown upwards and outwards leaving gaps and holes through which the cattle could squeeze their way with no trouble at all.

We spent an energetic day or two on the rews between the railway and the Downs, hacking and sawing down the unwanted bushes with billhooks and bowsaws and axes – a pleasant and warming job in the cold December days. Then we warmed ourselves even more, burning up the enormous piles of brushwood we'd cut – lifting and stacking the prickly bushes on to the biggest bonfire I'd ever seen with the ever-handy buckrake. "Hey, Fred," said the boss. "Whatever did we do before we had that tractor and buckrake? Can't seem to get along without it nowadays."

Fred was particular about the stems he wanted left on the old line of the hedge, making sure that there were plenty of long branches to lay over into the gaps. "'Tis no bliddy good fillin' in wi' dead wood what yer cuts from somewhere else," he said. "That'll rot away in no time, an' yer cattle'll walk through un fer a pastime."

Although we had no sheep on the farm, Fred told me that a hedge should always be laid strong enough to keep them in.

"If yer can 'old they boogers yer got a 'edge'll 'old anythin'."

"What do you mean, Fred? Surely sheep can't jump as high as cattle can? I would have thought that it was the other way round." Fred grinned at me, pleased to be able to show off his superior knowledge and have a dig at modern trends at one and the same time.

"Oh, ar – you got summat there, Boy. Cattle can jump, 'specially they gurt, leggy Fries'uns what 'e been getting' lately – jumps like bliddy deers they does – not like yer good old quiet Short'orns. But what I means is, sheep is real evil little devils fer workin' away at the bottom o' th' 'edge. If you han't got un good an' thick in th' bottom an' well plashed in, like, sooner or later one of 'em'll worm 'is way through, an' before yer can say 'knife', the rest on 'em's follered an' they're all in yer neighbour's corn."

Fred's hedges were good and well-laid – thick enough in the bottom to frustrate the wiliest sheep – and after a couple of years of careful trimming and grooming, high and dense enough to contain the worst of 'they gurt Fries'uns'.

I was disappointed that I wasn't allowed to stay with Fred and learn the fascinating art of hedge-laying, but the boss considered that it was more important to get the autumn ploughing finished, and I had my first proper lessons in this even more fascinating job – the most enjoyable on the farm. The boss had done all the ploughing on the farm when we still only had the one tractor, and my only experience was to relieve him during the lunch hour or if he was called away to deal with some emergency. That could hardly be called learning on the job: "Don't touch anything or alter the depth. Just keep the bloody thing straight and drop in and out of work at the headland mark." But with the coming of the Feguson that year, we now had two tractors. Of course, I was given the old Fordson and the two-furrow trailer plough, while the boss purred up and down on the much faster Ferguson, but that didn't detract from the enjoyment I got out of the job. And it was a good thing to learn the basic principles with a trailer plough which is more difficult to set and control than a modern, mounted one.

The main difference between trailer and mounted ploughs is that a trailer plough cannot be reversed. Very few fields are perfect squares or rectangles, so there is always a certain amount of 'short-work' – the places where the line of the ploughing is at an angle to the hedge, leaving a triangular place to be ploughed out. As it is totally against the rules of husbandry to turn the outfit on the ploughing –

leaving wheelmarks and breaking down the newly set-up furrows – you have to trundle up the headland running empty, until you can reach the other end of the bout and plough back to the short-work. Each bout of ploughing, or 'land', must be carefully marked out to minimise empty running which is wasteful of tractor time. With a mounted plough, though, when you reach the end of a piece of short-work, you merely lift the plough on the hydraulic lift, back up the furrow and then drop in again at the headland. All done in a matter of seconds and with very little planning or forethought needed.

A plough, which looks such an ungainly implement when it is standing in the yard or being towed along the road behind a tractor, changes its character completely when it sinks into the soil and starts doing the task for which it was designed. It seems more like some aquatic mammal than an inanimate machine, swimming through its natural element – cleaving its way along by skill and expertise rather than by force. The soil parts asunder before the cutting edges of the coulter and share, and inverts itself effortlessly over the curve of the mouldboard with a flowing, hypnotic motion that holds the driver completely engrossed.

A great part of the pleasure of ploughing is that one is always striving, like any artist doing creative work, for perfection – and perfection is never attainable. Is the rear furrow a fraction deeper than the front – just half a turn on the levelling handle needed? And is two notches of adjustment on the hake bar of the drawbar sufficient to counteract the sliding when we go across that little brow? Are the skimmers set correctly so that just enough green is pared off the side of the furrow-slice so there is no trace of it showing in the finished work? And while all this is going on the tractor must be steered, mostly by touch – a ploughman spends more time looking backwards than he does forwards – with the right-hand front wheel running along the furrow, just brushing the furrow-wall. The smallest kink in the furrow can turn into a mighty bow if it's not corrected almost before it appears.

Fred was naturally a master ploughman although, as he couldn't drive a tractor, he no longer did any ploughing on the farm. But there was still a single-furrow horse plough in the yard and Fred used to take this out sometimes at weekends. He used to earn himself quite a bit of money ploughing allotments and small pieces of ground that were too awkward for a tractor. Coe and Sons, the watercress growers, had a small, two-wheeled 'walking tractor', a British Anzani 6 hp job, which they used for ploughing their market garden ground, and when it was new everyone told Fred that his days of ploughing were over. But Fred said that he would wait and see "what sort o' job that there Tin Donkey would turn out". The driver of the Anzani 'Iron Horse' wasn't too pleased when he heard this insult but, as usual, Fred proved to be right. After a few attempts had left some of the allotments looking as if a snake had wriggled its way across them, scratching the soil in places, the expert was soon in demand again.

One of the first porkers in the yard died mysteriously about a fortnight before Christmas. The manner of its passing was no secret to us but, as meat was strictly rationed and killing pigs for home consumption was very nearly a capital offence, the affair had to be shrouded in a blanket of security.

I arrived at the yard one morning and was surprised to find the front gates and double doors leading to the pig yard closed and locked. Inside the yard, hidden from the road, was the village butcher's van. Clustered round – looking furtive – were the boss, Fred, and the butcher, who was carefully honing a keen edge onto a wicked-looking knife. I opened one of the doors and walked in.

"Get out!" they hissed at me. "And shut that ruddy door."

I stood and stared. "What's up? Has something died?"

"You will in a minute if you don't shut up." said the boss. "Go down to the front gate and stay there. Don't let anybody into the yard. And if that blasted policeman comes along, shout out, 'Help – the cows are out!"

"Are you killing a…?" The boss interrupted me fiercely.

"O' course not. You know it's against the law to kill pigs on the farm." He paused for a moment and then asked, "Do you know what a humane killer sounds like?"

"No," I replied, "I've never heard one go off."

"Good," he said, "because you're not going to hear one now. It's me shooting rats in the hen run if anyone wants to know, understand?"

I said I did and went to my post by the gate.

About an hour passed before the boss came down to the gate and said the coast was clear, the dark deed having been done and all traces of it removed. The only person who had passed by during that time had been the village postman. I told him he'd have to take the letters round by the garden entrance instead of his usual route past the pig yard and dairy. He asked me why. "Because the gates are all shut and I'm here – er – in case the cows get out."

The postman glanced at the blank doors of the pig yard and nodded understandingly.

"I sees what yer means, Boy. Killin' a pig, are they? You'll be all right tonight then, you will – fried chitterlin's fer supper, you'll 'ave." And he walked round by the garden whistling cheerfully, knowing that with a bit of luck – and a subtle, blackmailing hint or two about illicit pig-sticking – he'd be all right for a little bit of pork as a Christmas box.

Of course, I'd spent the previous hour in terror of the policeman coming – or worse still – one of the army of 'snoopers' who were government inspectors employed by the Ministry of Food to spy on villains who dealt in the 'Black Market'. They were loathed by everyone and likened to the Gestapo because of the methods they used.

There was one lovely tale of a farmer who telephoned his brother, also a farmer, to tell him that a snooper was in the district.

"Tryin' ter buy eggs at the backdoor, 'e is, off the ration, like. Told me a sad tale about 'is young wife bein' starvin' an' 'is kids bein' ill an' all. But I knowed 'im fer what 'e was – I seen 'im get out of a nice, shiny car just round the corner. If you sees 'im you set your dog on 'im." After a while the brother rang back to say that the snooper had visited him with the same tale of woe about starving kids.

"Felt sorry for 'im, I did. So I sold 'im a dozen eggs an' give 'em to 'im all done up nice in a egg box. Charged 'im three times black market price, I did, an' 'e paid up wi'out a murmer."

"You stupid idiot! An' after I warned yer, too. You'll 'ear more about that, you will!"

"That's what this snooper chap said," replied the brother, "but I don't reckon I will. There aint no law against sellin' china eggs at a quid a dozen is there?"

And if the village policeman had happened along the day the pig was killed I don't think there would have been any trouble. He called at the boss's mother's farm one day when they were getting ready to kill a pig. Bill, the tractor driver and general factotum, was a man who liked to dress for the occasion. He had on a butcher's apron and there was a steel hanging from his belt. He was just putting the finishing touches to his pig-sticking knife when the official knock came upon the back door. Bill opened it. There, on the doorstep, stood the tall, blue-clad figure.

"M-M-M-Mornin', Mr Bishop. I was just goin' to…" Bill's voice trailed away into silence. He couldn't think of anything legal he might be just about to do, dressed as he was. The policeman didn't bat an eyelid.

"That's all right, Bill, I only come round to sign your animal movement book." He looked thoughtfully at the knife, the steel, the apron and Bill's reddening features. "But I can see as 'ow you're real busy today. I'll come back another time."

The following Sunday the policeman and his wife had a nice piece of roast pork for their lunch.

The days leading up to Christmas were very busy. Fred and I worked flat out carting loads of hay, straw, mangolds and silage down to the yard so that we wouldn't have to do any carting on the two days of the holiday. All available trailers and wagons were loaded up and parked so that they could be taken out quickly to the field and their loads of fodder thrown off for the cows. The herd lived out of doors all the year round, now that we were using the outdoor bail, and it was much quicker feeding them that way than humping hay and silage into a cowstall. I had always heard that on the stroke of midnight on Christmas Eve, cattle tethered in a cowstall all went down on their knees. Now that ours were out in the fields I would never know whether they did or not. I asked Fred what he thought.

"I reckons they might," the old man replied.

"They does lots o' things we can't account for."

"But have you ever seen them do it?" I persisted.

"No, Boy, I han't." He looked at me warily, then saw that I wasn't going to tease him if he made an admission.

"I've allus wanted ter know, meself, all me life – but I han't never dared put it to the test."

The boss went off to Guilford one morning in his old Ford, to do his Christmas shopping. When he returned with his arms full of parcels he looked shocked.

"It's terrible!" he said. "I only got a few small things for the kids and I spent a fortune. Money goes nowhere these days. Do you know what this lot cost?" I had no idea but hazarded a guess and then doubled it, knowing that he was a bit of a spender when it came to presents for his two lovely little daughters, 'Manda and Cherry.

"Five quid?"

He snorted. "Over ten pounds – damn nearly eleven!"

I was astonished and impressed. Ten pounds was an enormous sum. It took me nearly a month to earn that, let alone spend it.

The evenings were busy, as well as the days, just before Christmas. There were dozens of cockerels, chickens, capons, ducks, and the occasional goose to be killed and plucked. I always hated the horrible job of plucking, which left my fingers sore and my hair full of poultry lice. But there was one consolation. The boss had a friend in the wine-import business. On Saturdays in the winter, when he went off to play rugger with the Dorking club, the boss sometimes took with him an empty gallon jar. When he returned, the jar was full of rum – not the mild stuff that is sold over the counter in pubs – but the full-strength, pre-blending liquor that had escaped by some devious means from the bonded warehouse. It was black and thick, like the West Indian molasses from which it had been distilled, and it had a kick like three mules. You couldn't drink it straight away at the start of an evening's plucking – you had to lead up to it with a few bottles of beer. But the boss used to watch the plucking carefully and directly we started to flag, out would come the gallon jar and there would be a small dose of 'the hard stuff' all round. Then things went with a swing and often the day ended with a spirited (literally) sing-song. Not, sad to say, Christmas carols, but songs that had come, with the rum, straight from the rugger club.

We had to milk a little earlier on Christmas Morning as the milk-lorry driver from Westcott Dairies had said that he would be making his collections early that day. He wanted to spend as much time as possible with his family, he said. So at half-past seven, the milk had been labelled, the calves and pigs had been fed and bedded-up, and there was only the outside feeding to do. The boss went off with the Ferguson to feed the heifers and Fred and I took the Fordson. We hitched up one of the loads of silage and I drove the tractor up to the top of the Hangers where the cows were. The cows clustered eagerly round the trailer, sniffing at the succulent, pickle-smelling load, and reaching up with their long tongues to try and steal a mouthful before it was unloaded. Fred climbed on the trailer and stood, ready to fork the silage off when I reached a clean piece of ground.

"Get off it, you owd varmint," he said as he tapped one

of the big, dewy noses with his prong handle. "You han't gettin' none 'til I gives it to yer."

I drove slowly along in first gear, while Fred pitched off heaps of silage onto the frosty grass. First one side of the trailer, then the other, with the heaps spaced out so that the cows couldn't reach each other with their horns while they were feeding.

"Spread it out, Boy, an' allus make more 'eaps than you got cows, then the little-uns 'll get a proper bellyful."

There is even a knack to feeding cattle out in the field. If you do it wrong the weaker animals in the herd don't get their fair share.

When the silage has all been thrown off the trailer Fred cleaned the last few wisps off the floorboards and then signalled me to halt. He climbed stiffly down and stuck the prong in the leg-shield in front of the Fordson's back axle. If it had been left on the flat trailer it might have been jolted off on the way home and got broken or lost. Then he walked back and strolled along the line of cows, inspecting them as they fed. I stopped the tractor engine and walked over and joined him.

"Don't look in bad nick, do 'em, Boy, considerin' as 'ow they 'as ter rough it livin' out?"

Coming from Fred this was high praise for the animals' condition. He wouldn't have dreamed of saying such a thing to Ernie, the cowman. There was a great deal of friendly rivalry between them and I had always thought that Fred was a bit put out when the cows became a one-man job and he had no more say in the feeding and milking.

"They look all right to me." I said. "If there was anything wrong I'm sure Ernie would have mentioned it when he brought the milk in."

"O' 'course 'e would 'ave," said Fred. "'e's a good bloke, Ernie is, for all that 'e do get a bit spiteful sometimes. But then – 'e's a cowman, in't 'e? All cowmen's like that."

"Why are you so interested in the cows this morning, Fred?" I asked.

"'Course I be a-feedin' 'em, Boy," was the simple answer.

"Allus likes ter take an interest in me work, I does. You do the same an' you'll be all right." The old man stopped and stood for a long time, admiring a newly-calved heifer that was quietly munching her way through a heap of silage.

"Winds it down 'em, don't 'em?" I loved that phrase and Fred used it often. It exactly describes how a cow reels in fodder, as if there were an endless conveyor belt joining her mouth and her rumen, down which fodder jogged in a continuous stream.

"They like their silage," I said. "But some of them took a couple of weeks to get used to it when we first started feeding it."

"Ar they are – that'll be them old cows what 'adn't never 'ad it afore. Bit partic'lar about changin' their grub, cows be." Fred took his wooden pipe out of his waistcoat pocket and filled it with black shag while he spoke. "I be the same. I likes me taters an' cabbage, an' if th' owd gel give me summat different I'd bliddy soon moan at 'er, I would."

"But you'll be having turkey with 'em today, won't you?" I asked. "It's Christmas Day. I see you've got your Sunday pipe on." Fred turned away from the wind and sucked the flame of a match down on the packed tobacco in his pipe. "Me 'ooden pipe's fer Sundays *an'* 'olidays, Boy," he said. "Clays is fer workin' days. 'Twouldn't be right to smoke a clay pipe on Chris'mus Day."

While I was wondering what bit of old Fred's obscure and private folklore forbade the smoking of clay pipes on high days and holidays, he reverted back to the subject of 'grub'.

"No, us won't be 'avin' no turkey fer our dinner today," he said. "In th' old days them as could afford it allus 'ad goose, an' us workin' chaps 'ad a bit o' pork off 've our own pig – an' then when times got a bit better like, a nice little roast o' beef. An' that's what me an' th' missus'll 'ave later on today – a bit o' beef. I tried that turkey once an' never did get on wi' it. Wishy-washy old white meat, that turkey meat – I likes a bit o' red meat what yer can get your teeth into." Fred suddenly jogged me in the ribs and pointed the stem of his pipe at the heifer. "Tell me about 'er, Boy."

I collected my thoughts. This was obviously one of Fred's little tests and I wanted to get it right.

"Her name's Jane. She calved about a fortnight ago and she had a bull calf. She's giving three and a half gallons a day and Ernie says he'll get her up to about four in a week or two." I couldn't think of anything else. "That good enough, Fred?"

"Do she kick when yer puts they machines on 'er?"

That made me think. As far as I knew Fred had never handled or milked that heifer and I wondered how he'd put his finger straight onto the only problem we'd had with her.

"She did to start with when she first calved. But she's not so bad now and Ernie says she'll be quiet enough when she settles down." Fred chuckled.

"Just like 'er mother, she is. You 'member, Boy, when you an' me was feedin' the 'eifers last Feb'ry, an' I told yer as that-un were old Jill's calf? I told yer she'd milk well, I did, an' I told yer she'd be a kicker, too."

"So you did, Fred, I remember now." He grunted approvingly.

"You 'member what us tells yer an' yer'll get on all right." he said. "You han't done so bad so far. Took us a year, it did, ter get yer straightened up, like, but yer don't get underfoot like yer used to, Boy."

I was pleased with the compliment but it would have been bad manners to acknowledge it. I changed the subject.

"Got round the feeding pretty quickly this morning, didn't we? Couldn't have done it so fast with horses."

Fred ignored this witticism and we walked back to the tractor and trailer. I swung the Fordson's starting handle lustily for a couple of minutes but the engine refused to start. Fred stood and grinned at me which didn't help.

"You go on, Fred," I said. "I'll have this thing going in a minute."

"Reckon I will, Boy." He turned to go and then paused. "An' if yer gets 'ome by mid-day, as it be Christmas, like, I'll buy yer a pint up at pub." He paused again, spat, and added, "Faster wi' tractors, in't it?"

Fred always had the last word.